Intermediate LENTIL SCIEN

I can set up, test, and write my own inequalities!

Science with Simple Things

Conceived and written
by **RON MARSON**

Edited and illustrated
by **PEG MARSON**

TOPS Learning Systems
342 S Plumas Street
Willows, CA 95988

www.topscience.org

WHAT CAN YOU COPY?

Dear Educator,

Please honor our copyright restrictions. We offer liberal options and guidelines below with the intention of balancing your needs with ours. When you buy these labs and use them for your own teaching, you sustain our work. If you "loan" or circulate copies to others without compensating TOPS, you squeeze us financially, and make it harder for our small non-profit to survive. Our well-being rests in your hands. Please help us keep our low-cost, creative lessons available to students everywhere. Thank you!

PURCHASE, ROYALTY and LICENSE OPTIONS

TEACHERS, HOMESCHOOLERS, LIBRARIES:

We do all we can to keep our prices low. Like any business, we have ongoing expenses to meet. We trust our users to observe the terms of our copyright restrictions. While we prefer that all users purchase their own TOPS labs, we accept that real-life situations sometimes call for flexibility.

Reselling, trading, or loaning our materials is prohibited unless one or both parties contribute an Honor System Royalty as fair compensation for value received. We suggest the following amounts – let your conscience be your guide.

HONOR SYSTEM ROYALTIES: If making copies from a library, or sharing copies with colleagues, please calculate their value at 50 cents per lesson, or 25 cents for homeschoolers. This contribution may be made at our website or by mail (addresses at the bottom of this page). Any additional tax-deductible contributions to make our ongoing work possible will be accepted gratefully and used well.

Please follow through promptly on your good intentions. Stay legal, and do the right thing.

SCHOOLS, DISTRICTS, and HOMESCHOOL CO-OPS:

PURCHASE Option: Order a book in quantities equal to the number of target classrooms or homes, and receive quantity discounts. If you order 5 books or downloads, for example, then you have unrestricted use of this curriculum for any 5 classrooms or families per year for the life of your institution or co-op.

2-9 copies of any title: 90% of current catalog price + shipping.

10+ copies of any title: 80% of current catalog price + shipping.

ROYALTY/LICENSE Option: Purchase just one book or download *plus* photocopy or printing rights for a designated number of classrooms or families. If you pay for 5 additional Licenses, for example, then you have purchased reproduction rights for an entire book or download edition for any **6** classrooms or families per year for the life of your institution or co-op.

1-9 Licenses: 70% of current catalog price per designated classroom or home.

10+ Licenses: 60% of current catalog price per designated classroom or home.

WORKSHOPS and TEACHER TRAINING PROGRAMS:

We are grateful to all of you who spread the word about TOPS. Please limit copies to only those lessons you will be using, and collect all copyrighted materials afterward. No take-home copies, please. Copies of copies are strictly prohibited.

ISBN 978 - 0 - 941008 - 52 - 5

= CONTENTS

PART I **Preparation** *and* **Support**

PART II **Reproducible Student Materials**

Watch children interact with their world, and you will see them grasp and poke and fiddle and manipulate. This happens nonstop, or so it seems, unless they are sleeping. What's going on?

You are witnessing, first hand, the serious and necessary work of developing human brains, doing what brains need to do in order to survive. These children are literally learning how to think, building and internalizing mental structures to facilitate abstract thought. Without first moving physical objects with their hands, children will never learn to move mental images in their heads. Elementary education in a nutshell is simply this: Kids need to *do stuff!*

"...self-organizing, autonomous learners need to do their own stuff to mature into well developed physical, emotional and intellectual human beings. "

Children are obviously doing stuff in today's schools. But is it the *right stuff*? Why do first graders, curious and ready for anything, too often turn into passive-aggressive seventh graders who fear math, hate science and misunderstand technology? In an effort to efficiently cover course requirements and meet state standards, well-intentioned educators have created a top-down, adult-centered approach to teaching that forces classrooms full of children to do the *same stuff* in the same way, as prescribed by adult lesson plans. This teacher-knows-best methodology may cover the syllabus, but it is not what students (especially younger students) want or need.

Education at its best happens on a knife edge between structure and freedom. Teachers know best when it comes to structuring and maintaining a safe, nurturing educational environment. But students also know best when it comes to making their own personal learning decisions. May we, as teachers, humbly remember that self-organizing, autonomous learners need to do their *own stuff* to mature into well-developed physical, emotional and intellectual human beings. Indeed, the most admirably self-actualized students become so because they are encouraged (and taught) to follow their own interests.

Teachers know how important it is to honor individual differences, to care for children at both ends of the ability spectrum as well as all those in the middle. But with so many children to teach, and so little time to do it, we sometimes falter. When class control begins to slip, we tend to revert to a safer top-down style of teaching that requires everybody to do the *teacher's stuff.*

This book invites you and your students to do a special kind of *TOPS stuff* with lentils. The Job Box provides a certain kind of *unstructured structure*. Its four walls define unambiguous limits. Yet, inside, the lentils remain fluid and malleable. An astonishing range and depth of concrete experience is possible in the comparing, searching, measuring, designing, dividing, calibrating and estimating of stuff.

Put a box of lentils in front of a child, and she will show you what she needs to learn. Throw in a few thoughtfully constructed manipulatives, and she will spontaneously develop her own course of academic study, from simple basics to sophisticated complexity! The brain knows *how* to develop. This wisdom is deep and intrinsic. Trust it. You can assist, but never control. This means that you teach to the child, and not the lesson. Let's see how this works:

▶ *Chris, which Job Card would you like to try today?*

▶ *OK. Go ahead and set up your Job Box. I'll return as soon as I can to see how you're doing.*

▶ *I see you're calibrating a liter bottle in cups. What volume reaches to this mark you've made here?*

▶ *May I pour in a pint of lentils to test this mark? No? Well, how do I know you put it in the right place?*

▶ *Oh, I see. You you're going to fill to your mark first, and then show me it fills the empty pint. OK...*

▶ *Sure enough! What if you filled to this mark again but poured it into this quart instead?...*

In typical interactions like these, the lentil box provides the structure. The student pursues his own interests within prescribed limits. And the teacher, through simple, direct questioning, exploits learning opportunities that pop up moment by moment. The student's mind is focused, ready to capture new insights, build new mental structures, grow smarter. Mental fusion over a box of lentils. A transfer of insight and knowledge, the *good stuff* of good education.

Testing the Waters: *Beginning with Confidence*

You are visiting Lentil Lake for the first time, and wonder whether you might go for a swim. First you dip your fingers and toes. If the water feels warm enough, you decide to wade in up to your waist. If the lake bottom feels firm and friendly, you take courage and decide to plunge in all the way. Next year, when you revisit these waters, you might decide to take a running dive. But for now, you want to test the waters of Lentil Lake step by step.

• *Fingers and Toes:*

Set up a single learning station, with one Job Box, in a corner of your room. Introduce the first Job Card to your entire class, using the materials indicated on the card. The first chapter is a nice place to begin, but you could start with another chapter just as well.

A single "person" icon on some Job Card headings indicates that the activity is best done individually. But two students *might* successfully work together, if they are good at sharing.

Students use this learning station, by permission, when their other work is complete. With many children and only one Job Box, working with lentils will be seen as a privileged activity, and students will be especially motivated to behave well. This might be a good time to introduce some of the Seven Easy Rules on page 13.

• *Waist Deep:*

When you are comfortable with the water temperature in Lentil Lake, **set up multiple learning stations**, with multiple Job Boxes. Each day or so, you might wade in a little deeper by adding another Job Box and/or a new Job Card. As you run out of dedicated space for learning stations, allow model students to retrieve materials from a designated storage area, then work at their own desks, on the floor, or wherever practical.

It is not necessary to do these chapters in order, or finish one chapter before moving on to the second. Quite the opposite. We strongly recommend that you introduce lessons horizontally, offering the first Job Card in every chapter first before introducing the second. This maximizes variety, providing students with rich developmental choice, and reduces the need to duplicate materials.

• *Take the Plunge:*

Are students becoming familiar with the rules? Do they handle equipment properly, clean up lentil spills and put everything back where it belongs? Are our recommended class procedures falling into place as you introduce them? Do students feel confident? Do you feel confident? Have you learned to trust the integrity and design of our program? Then you are ready to **mainstream the curriculum** into your class as a regularly scheduled science period.

No longer is our program supplementary, something to do at a learning station when other work is complete. Lentils have worked their way into your confidence (and into the corners of your classroom) as an important medium for learning. Now everybody does science together, in a thoughtfully structured program that honors diversity and maximizes free choice. Job boxes are scattered all around your classroom, with students working individually or in pairs.

And, even though this is *science* period, the *activities are so interdisciplinary*, your students are also learning math, language arts, social studies and art. Continue to work horizontally and vertically through the book. Feel free to suggest variations that occur to you, or substitute suitable materials. Improvisation is at the heart of creativity!

Within a few months, your students will be ranging far and wide through the curriculum, happily engaged in designing their own projects, initiating deeper investigations into their areas of greatest interest, moving beyond the Job Box. Ah, yes. The water in Lentil Lake is really fine.

GLOSSARY *of Basic Materials and Procedures*

All supplies used in more than one chapter, and instructions for turning them into manipulatives for student use, are listed below. All tools needed to make any item in this book are also listed here. Items dedicated for use in a single chapter can be found on a special materials page in that chapter. Any time you find an item underlined, here or elsewhere, look for a more detailed explanation in this glossary.

Some items are needed in quantity. We begin these descriptions with two numbers separated by a dash and ending with a colon, *i.e.* 2–4: . The first number estimates what you'll need for a single learning station set up in the corner of your classroom or at home. The second estimates what you'll need to accommodate 30 students working alone or in pairs during a dedicated science period.

baby food jars (BFJ's)

2–4 sets of 3 baby food jars: A trio of small (2½ oz), medium (4 oz) and large (6 oz) baby food jars. Gerber brand calls these 1st, 2nd and 3rd foods, respectively. Masking tape labels: *small, medium, large.*

booklets

1. Photocopy the associated line master. (See page 23 for an example.) Fold it along the grey center line.

2. Trim off the long border, parallel to the fold, cutting through both layers of paper.

3. Cut *toward* the fold (to prevent fanning) on the dashed lines between the 8 folded pages. Discard the trim.

4. Order the pages, folded ends to the right, with page 1 on top. Page numbers will be in the upper right corner.

5. Crease the folded edges of the collated pages. Jog them even, then bind the uneven cut edges on the left with a single staple.

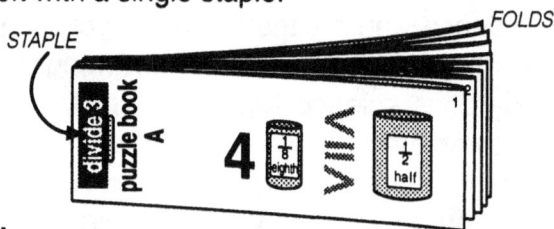

bottle caps

31–34: Use twist-off caps that are not bent. If beer labels are offensive, cover with masking tape.

bottle-cap spoons

1–4: Fix a bottle cap to the end of a craft stick with rolled masking tape.

bottle lids

5–50: These are especially important for sealing lentils in liter storage bottles away from moths, mice and other pests. All bottle lids must measure 1⅛ inch (3 cm) in diameter. Some manufactures make bottles with wider mouths. Avoid these completely. When not sealing containers, store lids in a container of appropriate size and label: *bottle lids.*

box and brick

A thrifty filing cabinet. Use to store student folders.

cleaning screen

Use ⅛ inch mesh hardware cloth (wire screen). Frame a 6 inch square (or somewhat larger piece) in electrical tape to cover wire ends. Bend up the perimeter on 3 sides to form a shallow dust-pan shape. Use this to clean lentils swept off the floor. Shake the screen over a waste basket to filter out dust and dirt. Pick out larger refuse by hand. If swept-up spills are dumped directly into the Job Box, your lentil supply will grow dirty and trashy over time. (*NOTE:* When purchasing this material, get some ¼ inch mesh as well. See *screen*, page 9.)

cleanup

Use any procedure that works for you. Here are some options. ✓Lentils are big enough to pick up off the floor in moderate numbers. These are still clean enough to return straight to the Job Box. ✓Sweep larger spills into a cleaning screen, filter and return to the Job Box. ✓Provide a wide-mouth holding container labeled "dirty lentils." Direct students to return all swept-up lentils here for later screening by a student helper. ✓Discard swept or vacuumed lentils, replacing as needed. Lentils are, after all, quite inexpensive and totally biodegradable.

clear cups

12–20: These are tall, tapered, clear plastic, 10 ounce cups used both in lentil activities and as storage containers. Solo and Polar brands are suitable.

clear tape

1–3 rolls: Use ½ inch or ¾ inch wide tape.

craft sticks

40–100: Also called popsicle sticks. Available by the box at reasonable prices from craft stores. Store in a clear cup.

cups

When we refer to a cup (or fractional cup), we mean amber pill vials (often cut to a particular size). See also standard cups, cut-to-size containers, and plastic vials.

cut-to-size containers

All volume measure in this book is based on 60-dram plastic vials, commonly used to package prescription drugs, and also available from TOPS. Even though 60 drams is officially equivalent to only 94% of a real cup, it's close enough to look and feel like a real cup. More important, all our related measures (pints, quarts, half cups, etc.) derive from this standard.

And how will you derive them? Some containers can be used as manufactured. Other need to be cut to size, using templates we provide. And others can be sized experimentally by pouring lentils.

Suppose, for example, you need to create a pint that holds precisely 2 standard cups. Follow the instructions below. (Then generalize this procedure to derive other standards you need to make.)

1. Overfill a standard cup with lentils. Shake once to shed excess lentils. You now have a slightly rounded, loosely packed cup that we call "fair and full." Funnel this cup into a half-liter bottle. Add a second cup in the same manner. Tilt the bottle, if necessary, to level the contents, but don't shake it down.

2. Mark the level of the lentils with a dot from a permanent marker on the outer surface. (Note: If your container is translucent, hold it at eye level against strong light to see the lentils inside. If the container is opaque, bring your fingers together inside and outside, level with the surface of the lentils. (It's OK if this reference dot is only approximate.)

3. Hold the marker absolutely steady against your reference dot by bracing your hand on a stationary object of suitable height. A pile of books works well.

4. Rotate the container against the steady marker tip with your free hand, inscribing a level line around the container, back to your starting dot. (Note: If the container is square, a milk carton perhaps, slide it past the marker on all four sides.)

5. Inscribe two more lines around the container, one a little above your reference dot, and another a little below. (These extra lines are not necessary if your original line is certain.)

6. Cut around the top line with curved toenail scissors. Test your pint at this height by pouring in 2 standard cups filled fair and full. Cautiously trim parallel to lower lines as necessary, until the level is precise. (Note: If the container is small and curved, you may wish to spiral the cut downward from the top, converging with your inscribed line. Always hold the scissors so the blades curve *away* from the line you are approaching.

7. Always feel for sharp edges and snags. If the edge remains too sharp, cover with masking tape. (We have never found sharpness to be a problem.)

electrical tape

Use a roll of $3/4$ inch black plastic tape. This is a teacher supply used in small quantities to prepare student materials.

extra cups jug

1–2: Whenever a Job Sheet specifies "extra" cups, students should retrieve plastic vials from this jug, leaving *dedicated* cup sets related to other chapters (Compare, Measure and Divide) intact for other Job Cards.

Prepare a gallon storage jug to hold 6 whole cups plus fractional cups as specified on page 108. See page 81 for cutting each part to size. Fix a paper clip to the side of each whole cup (6 in all) as follows:

1. Roll masking tape into a small tube, sticky side out. Center it, with open ends aimed up and down, to the side of the cup.

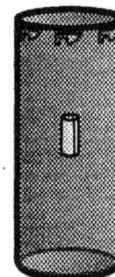

2. Stick a new (not bent) paper clip on the tape as illustrated. The ends should point down, with the double loop projecting just above the edge of the tape.

3. Tape tightly over the top, matching the top edge of the rolled tape underneath.

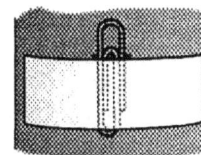

film cans

4–8: Use plastic film canisters with snap-on lids.

floor cleaning equipment

Use any of the following: a broom, whisk broom, dust pan, cleaning screen, carpet sweeper or vacuum.

funnels

1–20: Cut these from one-liter or two-liter bottles. Because you'll need as many as 40 matching liter bottles for lentil storage, have them in hand before cutting up extras for funnels. Cut mismatched or odd-shaped bottles first. Or make funnels from 2-liter bottles, which are used for little else in this curriculum.

Directions for cutting bottles of either size are the same: Measure down the sloping shoulder about 2 inches from the base of the neck, and make a dot. Inscribe a circle at that level as in cut-to-size. Weight the bottle with lentils or water, if needed, to help steady it as you turn it. Pierce the bottle with pointed toenail scissors and cut along the inscribed line. Your cut will be somewhat ragged because of geometric constraints. Trim as necessary to make it even. Feel for sharp snags. Funnels made from either bottle stand equally tall, measure about 9 cm in diameter, and funnel nearly equal capacities.

gallon storage jugs

5-13: Cut the widest possible opening in the top of each jug with toenail scissors. Keep the handle intact.

hacksaw

Needed for material preparation only in D/ Design.

half-liter bottles

2-6: These are found in grocery stores and convenience store cold cases. The best shape is tall and straight. Ribbed sides are OK. Bottles should look identical. If the screw-on lids also fit the 1-liter bottles, this is ideal.

Masking tape labels: *half liter*

index cards

14-14: Use 4 x 6 inch cards.

job boxes

1-20: We used 19 x 14.5 x 3 inch (49 x 37 x 8 cm) corrugated cardboard boxes, commonly supplied by nurseries to carry home potted plants. These are cut and scored to fold into self-locking boxes, with seamless bottoms that are especially good for scooping and pouring lentils. You might negotiate a special price for a bundle of unfolded boxes, then assemble them with your students as a class project.

Cover all inside seams – anywhere lentils might stick or hide – with clear packaging tape. Keep the bottom entirely free of tape. (It tends to peel up over time, as students scoop up lentils.)

Plastic tubs or sorting trays with similar dimensions also work.

Add a holder for the Job Cards: Cut a strip of packaging tape about ¾ the length of one of the longer sides. Center it inside the box on this longer side, flush with the bottom edge. Cut a 6 inch (15 cm) length of drinking straw. Fix it horizontally on the taped side, about a finger width from the bottom of the box, with masking tape at each end.

Test to see that a folded Job Card slips easily into this holder.

JOB CARD

lab coats and ID badges

Recycle white shirts and photocopy ID badges as outlined on page 15. These props are optional for establishing alter-ego famous-scientist identities.

lentils

4-80 pounds: Say what? No, this is not a typo. About 20 Job Boxes, on average, are used by a class of 30 students working alone or in pairs. Each Job Box typically requires 1 or 2 liters of lentils to support Job Card activities, and each liter bottle holds nearly 2 pounds.

I can see you now, dear reader, closing your eyes and imagining how scary it might feel to be stuck in a classroom with 30 youngsters and 80 pounds of lentils. Please see *Testing the Waters* on page 5 to find our comforting suggestions for introducing **Lentil Science** at *your own* pace.

Eighty pounds of lentils seems strange only because modern classrooms are generally bereft of volumetric experiences. Except for occasional rice tables on the K-1 level and demonstrations with water at higher levels, children have very few opportunities to experience the qualitative and quantitative character of 3-dimensional space. **Lentil Science** addresses this problem in a thoughtful, organized way. You *can* manage it. And your kids will be *way* smarter for it!

Find lentils, sold in 1 or 2 pound bags, near the rice and beans at your local grocery store. Or purchase them in bulk at wholesale food outlets. Or strike a special high-volume discount deal with your grocer. Or ask for help from family members (see page 10 for a letter you can send).

Always store lentils in closed bottles when not in use, away from moths, mice, and other pests.

liter bottles

5-50: Collect both plastic bottles and bottle lids. Most will be used for lentil storage, some for Job Card experiments. These commonly package ginger ale, seltzer water, and other carbonated drinks appealing to adult tastes. Try to collect a matching set of bottles with equal height and width. Simple, smooth-sided cylindrical shapes work best. Reject ribbed or unusually-shaped bottles unless nothing else is available. Odd liter bottles can be cut into funnels.

If you can't locate enough one-liter bottles right away, substitute the more ubiquitous two-liter soft drink bottle as temporary holding containers. These larger bottles are probably too flimsy to hold up over the long term, and somewhat awkward to use.

magnets

4-16: Rectangular ceramic refrigerator magnets. See chapter B/Search, page 33.

manila file folders

8-37: Use folders sized to hold 8½ x 11 inch paper. These organize curriculum materials and student work.

masking tape
1-5 rolls: Use standard $3/4$ inch wide tape.

packaging tape
Use clear tape on a 2-inch-wide roll. Opaque tape is *not* a good alternative. This is a teacher supply used widely to prepare student materials.

paper clips
Use one box of standard sized clips.

paper punch
Teacher use only.

permanent marker
Teacher use only. Select a fine point with black ink. "Sharpie" brand works well.

plastic vials
We used pill vials manufactured by Owens-Indiana, with an "O-I" stamped on the bottom, along with the letter T, followed by a number indicating its volume in drams. (See page 10 for an illustration.) The tops are rimmed by "child-proofing" nubs. We selected these because of their wide availability in national drug store chains. You can also purchase them directly from TOPS in all needed sizes.

The first quantity given for each item (for a single learning station) includes enough vials to produce one complete set of materials. These include milk-jug container sets for Compare, Measure and Divide, an extra cups jug set, plus a few additional vials (each dedicated to a special use).

The second quantity given (for a class of 30 students) includes one set of everything above, plus additional sets of vials for Compare, Measure and Divide:

10-22:	60 drams
3-9:	40 drams
2-5:	30 drams
2-4:	20 drams
5-15:	16 drams
5-14:	13 drams
5-19:	$8^1/_2$ drams

scissors
A pair of adult scissors for teacher use, and safety scissors for student use.

scoops
1-20: Use quart or pint cardboard milk cartons. (Half pint school-lunch cartons make an acceptable, though somewhat smaller scoop.) Cut to size: cut off the top with toenail scissors, leaving a box about $1^1/_2$ inch (4 cm) tall. Cut away one of the sides. (Alternatively, you might leave all four sides intact. A box scoops more efficiently and pours well, but doesn't have the classic shape.)

This important tool is always needed for clean up, to return lentils to storage bottles. It is only specified on a Job Card however, when used as an integral part of the investigation.

screen
Use $1/4$ inch mesh hardware cloth (wire screen). Needed to make sorting screens in B/Search, and counting grids in G/Estimate.

standard cup
10-22: This terms refers to 60 dram Owens-Indiana plastic vials rimmed with lock-tight nubs. Their volume is the standard by which all other measuring containers in this book are calibrated.

stapler
One on the teacher's desk is enough.

stick rulers
1-4 sets: Photocopy page 51 and follow the instructions.

straws
1-20: plastic drinking straws. Used to hold Job Cards in Job Boxes.

toenail scissors
These must be curved and pointed. They are used by the teacher for cutting various plastic and cardboard containers to size. Heavy-duty scissors work best.

tubs
3-9 Try to locate a uniform set of 1-pound margarine or butter tubs. These generally hold 3 standard cups, as manufactured. This size is ideal; a slightly larger size is OK. Do NOT substitute smaller sizes.

white glue
Used by teachers to prepare student materials. Apply from bottle or stick. Student use is optional, an alternative to tape.

wire cutters or shears
Needed to cut hardware cloth to size in B/Search and G/Estimate.

wood saw
Needed to cut two 2 x 4's to size in B/Search.

Dear Family Members of _____,

    We are setting up a hands-on learning program called **Lentil Science** *at our school. We will be using lots of lentils, bottles, and pill vials, to pour, search, compare, design, measure, divide and calibrate. Any help you might give us in purchasing lentils or gathering the recycled materials listed below is much appreciated. We need to collect most of these materials by* _____.

Sincerely,

QUANTITY	
Program requires:	We currently need:
20	

corrugated cardboard boxes with seamless bottoms, measuring approximately 19 x 14.5 x 3 inches. (49 x 37 x 8 cm). These are commonly supplied by nurseries to carry home potted plants.

80	

pounds of lentils in bags of 1 pound or more.

70	

clean, dry, one-liter plastic bottles, with **screw-on lids** that measure $1\frac{1}{8}$ inch (3 cm) in diameter. These commonly package ginger ale, seltzer water, and other carbonated drinks appealing to adult tastes. Simple, smooth-sided cylindrical shapes work best. Please don't confuse this 1.0 liter size with larger 1.5 liter and 2.0 liter sizes also on the market.

20	

one-quart milk cartons made from cardboard, not plastic.

13	

one-gallon milk jugs, plastic.

30	

old white dress shirts to recycle as lab coats.

plastic pill vials, amber color with child-proof "nubs" around the rim. We don't need the lids. The brand we need is manufactured by Owens-Indiana. These have an "O-I" on the bottom, along with the letter "T" followed by a number showing its volume in drams. We are looking for the following sizes, listed from largest to smallest:

Program requires:	We currently need:	
22		**T-60 pill vials**
9		**T-40 pill vials**
5		**T-30 pill vials**
4		**T-20 pill vials**
15		**T-16 pill vials**
14		**T-13 pill vials**
19		**T-8$\frac{1}{2}$ pill vials.**

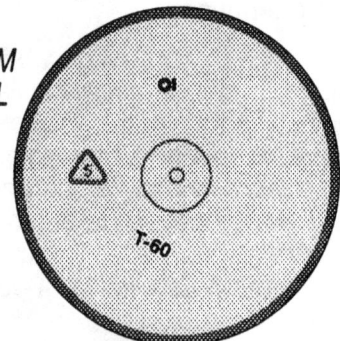

BOTTOM OF PILL VIAL

How to Organize Work Space *and* Materials

1. Prepare 58 Job Cards: Make one photocopy of each card. Shift the book on the copy window until you find the best position for a reasonably well-centered image. Fold each photocopy twice, first bisecting the *horizontal* gray line, then the *vertical* grey line. This creates a "greeting card" format, with student directions on the front of each card, needed materials on the back, and teaching notes handy on the inside. (Card corners and edges may not match as a result of imperfect centering. You may trim these edges even, if you wish, but no one is likely to notice.)

2. Prepare 7 Chapter Folders: Center the long side of a 4 x 6 inch index card along the spine of a standard-sized manila folder. Tape it with wide packaging tape around this spine, and up to the top of each short side. Leave the last long side open to form a pocket. Slip a chapter set of Job Cards into the pocket, and Job Sheets into the folder. Prepare 6 more Chapter Folders in the same manner, and label the tabs.

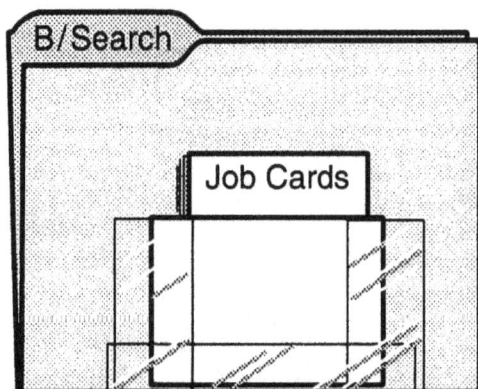

3. Photocopy Job Sheets: These pages follow some Job Cards to support written work. The label *Job Sheet* appears in the lower right corner of the page, followed by suggested quantities to photocopy (usually 1 per student). Paper clip duplicated sets together, and store in numerical order inside corresponding Chapter Folders.

Note: Photocopy only a few of each Job Sheet to start, until you develop a sense of how fast they are being consumed. You can always copy more later.

4. Prepare Student Folders: Assign each student a class number corresponding to the position of each name on your finalized class list. Boldly label the tab on a standard-sized manila folder with each student's assigned number to facilitate quick and easy filing.

Photocopy class sets of pages 12 and 14. Staple the *Seven Easy Rules* and *Progress Chart* inside the front cover. Fold the *Borrowing Cards* in quarters, then paper clip one inside each folder.

At a later time you might have students write their names (real, or assumed *Famous Scientist* names) on the covers of their folders, and personalize them with decorative science themes.

5. Prepare Special Materials: Find detailed instructions at the beginning of each chapter for gathering and assembling these items. Photocopy each page of associated labels to tag special items. Use dayglow or brightly colored copy paper if available. Store all chapter materials together near each chapter sign. (These require 1 square foot of space, or less, per chapter.)

6. Prepare Basic Materials: Send the *Dear Family Member* letter home with each child (see opposite page) to enlist outside help in gathering *high-quantity basics*. Store these together, on and under a table or counter:

> Job Boxes
> liters of lentils
> bottle lids (store in a labeled container)
> scoops
> funnels

Store *low-quantity basics* together, near a stand-up sign or in a labeled box that reads "BASICS." Photocopy this sign on page 45 and gather the indicated materials. Apply the label(s) on page 108.

7. Organize Cleaning Materials: You might provide *some* of these in a corner of your room, depending on your <u>cleanup</u> strategy:

> <u>cleaning screens</u>
> <u>floor cleaning equipment</u>: broom, whisk broom, dust pan, carpet sweeper or vacuum
> a "dirty lentils" container

8. Issue <u>Lab Coats and ID Badges</u>. These are optional, for classes using alter-ego identities of *Famous Scientists* (see page 15). Medium brown paper bags, labeled by class number, may help you organize coats and badges.

Seven Easy Rules

1. Be safe.
• Never throw lentils, or anything else.
• Never put lentils in your mouth, ears or nose. (Lentils in ears may need a doctor to remove.)

2. Respect others.
• Imagine you are in a bubble as big as your arms reach. This is your personal space. Don't enter another person's bubble unless they invite you.
• No put-downs or name-calling.

3. Avoid spills. Clean them up.
• Work inside your Job Box.
• Clean up all spills as soon as they happen.
• You may tip a Job Box to gather lentils, but it should always touch your table. No Job Box may be carried around unless it is empty.

4. Stay on task.
• Display the Job Card you are working on.
• Use only the materials pictured on the Job Card. (You will often need to return empty liter bottles and lids to storage. Sometimes the scoops and funnels should stay in storage, too. Get them when you are ready to clean up.)

• If you want to do your own experiment, display Job Cards that say "On your own."

5. Finish what you start.
• Finish old work, and ask the teacher to check it, before starting a new Job Card.
• You have "first dibs" when continuing a Job Card the next day, as long as you don't keep it too long.

6. Take care of equipment.
• Put everything back where it belongs, so everyone can find what they need next time.
• Keep lentil storage bottles closed.
• Don't keep equipment for yourself. It belongs to all of us. When it's gone, everyone loses.

7. Eat a "balanced diet."
• Work on many different Job Cards, both those you love to do **and** those you need to learn.
• Each time your teacher OK's a finished Job Card, fill in *one* white line in the correct box on the Progress Chart below. Each box holds up to 4 marks. Can you earn a mark in every column? In every box? That's good balance!

Progress Chart for _____

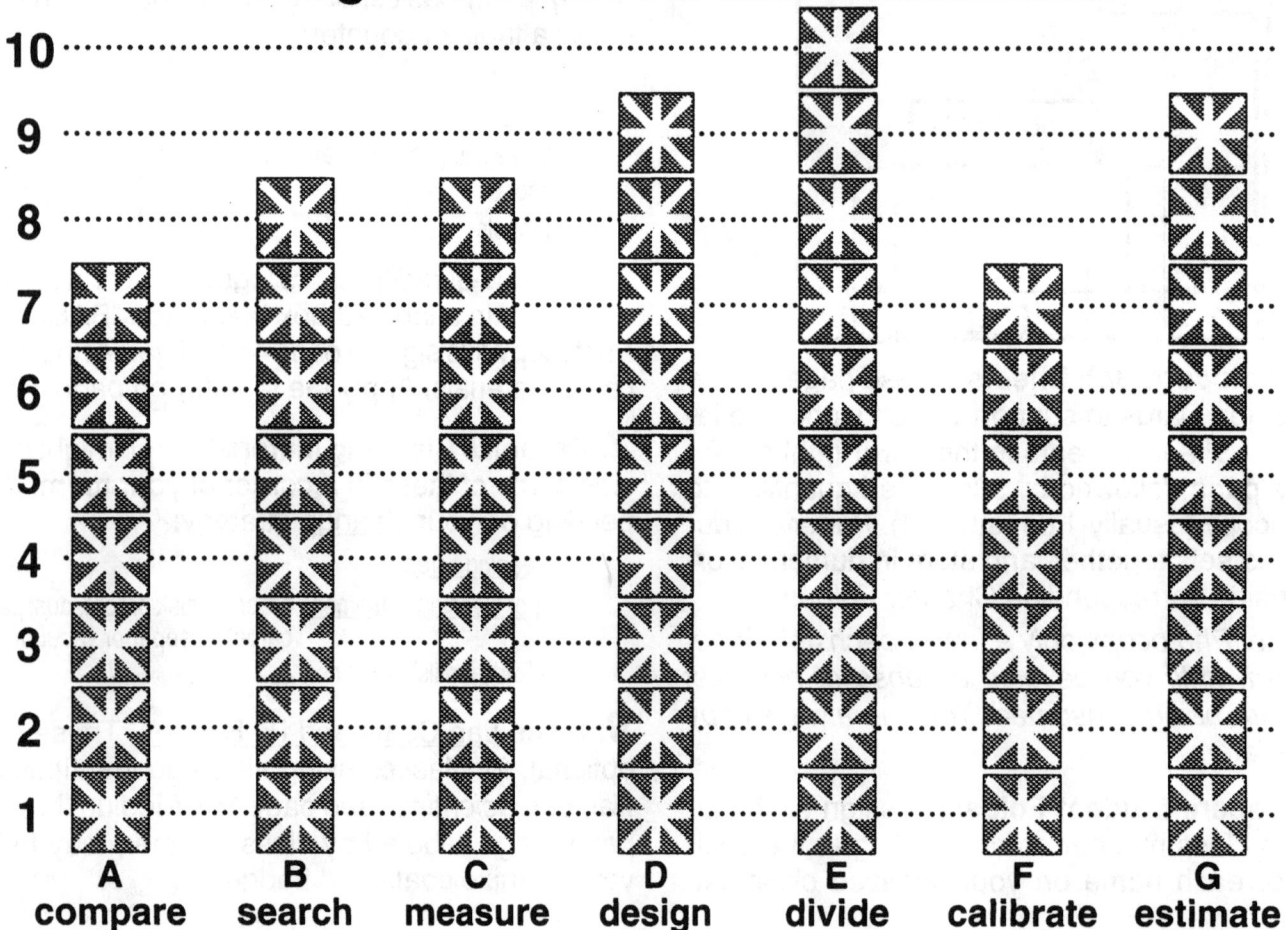

| | A compare | B search | C measure | D design | E divide | F calibrate | G estimate |

Model these steps for your students, then ask them to role play until you're sure they've got it. Review if necessary.

Every science period:

1. Look inside your Student Folder. You'll find Seven Easy Rules, a Progress Chart, and a Borrowing Card with a calendar on it. This is where you'll put all your written work.

2. Take your Borrowing Card to the Chapter Folders, and choose a Job Card to work on. (If you haven't finished work on your last Job Card, or didn't get a teacher check, go back to that one.)

Note: Allow students who wish to continue with an unfinished Job Card to choose first, as long as they're not monopolizing a popular card. All remaining cards then become "free" for others to borrow. (Only those cards that you have already introduced should be in circulation. Keep adding new ones day by day.)

3. Borrow a Job Card:
• Write its *Letter/Number* in today's box on your Borrowing Card.
• Put your Borrowing Card in the pocket of the Chapter Folder.

Note: Borrowing allows you to see at a glance who is using which card. And Job Cards are less likely to get lost if students must return them to get their Borrowing Cards back.

4. Find out how many people should work on that Job Card. Little "person" symbols on the front of each folded card show if you should work alone or with a partner. If you find someone to work with, put their Borrowing Card in the Chapter Folder pocket along with yours.

5. Gather materials. Bring everything you see on the back of the Job Card to your work area. Each item has its own special place. If Job Sheets are shown, find these in the Chapter Folder. Look for all materials near the chapter sign (or in the chapter box) and near the "basic" sign (or in the "basic" box). Finally, get a Job Box and however many bottles of lentils your Job Card shows.

6. Stand the Job Card in the holder at the back of the Job Box. Pour in lentils if the Job Card tells you to. Go to work. Experiment. Ask for a teacher check any time you're ready.

7. Ask for a checkpoint when you finish your work. When your completed work has been OK'd, mark a line on your Progress Chart. (Colored pencil looks nice.) If you have time to start a new Job Card, trade the old one back for your Borrowing Card and find a new one to record and swap. (Squeeze the extra number into today's box, or use tomorrow's box.) Return old materials and gather new materials as required.

Note: Initial your approval at the top of each student's work. This might be a completed Job Sheet, a story, answered questions, a drawing, a computation, just about anything. Even very young children can produce some kind of recorded response. At the very least, they can copy the number of the card they have been working on for your initialed approval. Gently nudge each individual toward a higher personal standard, greater elaboration, more refinement as they grow and develop.

8. Follow the rules stapled in your Student Folder. Be nice. Practice care and respect for each other and for the equipment.

Note: You can reduce theft in your classroom by instilling a sense of group ownership. Introduce equipment with care and reverence. Emphasize how you made each new item or where you purchased it, how it works, and where it belongs. Then explicitly gift it to all class members: I give this to you, and you, and you..., to help you learn and grow.

A "balanced diet" is also very important. Respect the deep wisdom of each child's mind and body to seek nourishment in its own natural way. But watch out for those children who have already become "sugar junkies," looking for fun, taking the easy way out. Use your adult perspective and wisdom (and stickers and stars!) to gently coax these children to work broadly through all subject areas, earning many marks across their Progress Charts. As they do this, you'll enjoy the added benefit of fewer materials serving more students. Everybody wins!

9. Please clean up when I tell you to. Be careful, courteous, quiet, and quick. Put everything back where it belongs. Return the Job Card to the correct Chapter Folder pocket, and take back your Borrowing Card. Put *all* of your papers, including Job Sheets, artwork, stories and such, in your folder.

Borrowing Card

name: my class number:

SEPTEMBER

Day:	S	M	T	W	T	F	S
Week: 1							
2							
3							
4							
5							

OCTOBER

Day:	S	M	T	W	T	F	S
Week: 1							
2							
3							
4							
5							

NOVEMBER

Day:	S	M	T	W	T	F	S
Week: 1							
2							
3							
4							
5							

Borrowing Card

name: my class number:

DECEMBER

Day:	S	M	T	W	T	F	S
Week: 1							
2							
3							
4							
5							

JANUARY

Day:	S	M	T	W	T	F	S
Week: 1							
2							
3							
4							
5							

FEBRUARY

Day:	S	M	T	W	T	F	S
Week: 1							
2							
3							
4							
5							

Borrowing Card

name: my class number:

MARCH

Day:	S	M	T	W	T	F	S
Week: 1							
2							
3							
4							
5							

APRIL

Day:	S	M	T	W	T	F	S
Week: 1							
2							
3							
4							
5							

MAY

Day:	S	M	T	W	T	F	S
Week: 1							
2							
3							
4							
5							

Borrowing Card

name: my class number:

JUNE

Day:	S	M	T	W	T	F	S
Week: 1							
2							
3							
4							
5							

JULY

Day:	S	M	T	W	T	F	S
Week: 1							
2							
3							
4							
5							

AUGUST

Day:	S	M	T	W	T	F	S
Week: 1							
2							
3							
4							
5							

Fill Your Class with Famous Scientists

Ego often gets in the way of learning. We all know this. We see it in our students and feel it in ourselves. We *want* to learn new things, of course. But first we have to protect our self-image – make certain that our ignorance, real or imagined, is never discovered.

Lozanov, a Bulgarian psychologist and language teacher, called this problem a learning barrier. He found that he could sidestep problems of ego by allowing his language students to role-play new identities. You can, too. Try this in your class:

1. Photocopy the names of famous scientists on the next page. Read some of the famous names from this list to your class. Post a copy on your bulletin board or hand out duplicates. Ask students to think about who they might like to be.

2. Ask each student to choose a name from this list. Girls can cross the gender gap by adopting a feminine version of any male name. Two or more students might choose the same name: *Issac Newton I* and *Isabel Newton II*, for example. Some may wish to take the names of more recent scientists not on the list: Carl (Carla) Sagan, Linus (Lucy) Pauling and many others.

3. Photocopy, cut out and distribute the ID badges. Students should fold them in half, neatly fill in both sides, then sign the bottom. Laminate these tags (or cover front and back with clear packaging tape), paper punch, and provide heavy string or yarn for wearing around the neck.

4. Ask each student to introduce his or her scientist-self to the class and talk briefly about background and accomplishments. You might assign students to research their "autobiographies" in the library or on the internet. At the end of each presentation, students might hold up their signed card as evidence that they have "taken the pledge."

5. Issue lab coats (old white shirts) for students to wear with their name tags.

---

first name

last name

achievements

My pledge:

When I wear this name tag...

1. I will pretend to be the scientist I have chosen, in a manner that honors their memory.

2. I will address my fellow scientists by last name (Sir Newton or Madam Newton or Colleague Newton), and treat them with respect.

3. I will observe carefully, think critically and report accurately, to the best of my ability.

X_____
(science name)

X_____
(real name)

---

first name

last name

achievements

My pledge:

When I wear this name tag...

1. I will pretend to be the scientist I have chosen, in a manner that honors their memory.

2. I will address my fellow scientists by last name (Sir Newton or Madam Newton or Colleague Newton), and treat them with respect.

3. I will observe carefully, think critically and report accurately, to the best of my ability.

X_____
(science name)

X_____
(real name)

Support: 1 copy per 2 students

Howard (Holly) **ALKEN**: developed first large-scale digital computer

Amedeo (Amedea) **AVOGADRO**: determined Avogadro's number of atoms in a mole

Alexander (Alexandra) Graham **BELL**: patented the telephone

Daniel (Daniela) **BERNOULLI**: studied fluid movement and pressure; explained how airplanes fly.

Henry (Henrietta) **BESSEMER**: developed a process for steel production

Niels (Nielsa) **BOHR**: leading developer of the quantum theory

Robert (Roberta) **BUNSEN**: developed the Bunsen burner

George (Georgia) Washington **CARVER**: agricultural chemist

Henry (Henrietta) **CAVENDISH**: discovered hydrogen

Madam (Sir) **CURIE**: noted for work with radium

John (Joan) **DALTON**: developed the atomic theory

Charles (Charlotte) **DARWIN**: developed the theory of organic evolution

Thomas (Tamara) **DOOLEY**: jungle doctor

Rudolf (Rhoda) **DIESEL**: patented the Diesel engine

Christian (Christine) **DOPPLER**: demonstrated the Doppler effect

Thomas (Thomasina) **EDISON**: inventor of the electric light bulb and phonograph

Albert (Alberta) **EINSTEIN**: developed the theory of relativity

Leonhard (Leonora) **EULER**: authored the first calculus book

Gabriel (Gabreille) **FAHRENHEIT**: introduced the Fahrenheit temperature scale

Michael (Michelle) **FARADAY**: noted for work with electricity

Alexander (Alexandra) **FLEMING**: discovered penicillin

Sigmund (Sigrid) **FREUD**: founder of psychoanalysis

GALILEO: founder of the experimental method

Robert (Rhonda) **GODDARD**: founder of modern rocketry

William (Wilma) **HARVEY**: discovered circulation of the blood

Edwin (Edwina) **HUBBLE**: discovered evidence of the expanding universe

Julian (Julie) **HUXLEY**: noted philosopher of science

Edward (Edna) **JENNER**: vaccination pioneer

Carl (Carla) **JUNG**: pioneer in analytical psychology

Sister Elizabeth (Brother Ellis) **KENNY**: developed treatment of polio

Johannes (Johanna) **KEPLER**: developed laws of planetary motion

Antoine (Antoinette) **LAVOISIER**: founder of modern chemistry

Louis (Louise) **LEAKEY**: discovered fossil remains of early hominids

James (Jane) Clerk **MAXWELL**: developed Maxwell equations of electromagnetism

Maria (Marion) Goeppert **MAYER**: discovered structure of the atomic nucleus

Gregor (Greta) **MENDEL**: discovered heredity, dominant and recessive genes

Albert (Alberta) **MICHELSON**: established the speed of light as a constant

Robert (Roberta) **MILLIKAN**: investigated electronic charges and the photoelectric effect

Thomas (Tamara) Hunt **MORGAN**: developed the chromosome theory of heredity

Isaac (Isabel) **NEWTON**: discovered laws of gravitation and motion

J. Robert (Robin) **OPPENHEIMER**: developed the atomic bomb

Louis (Louisa) **PASTEUR**: developed the process of pasteurization

Max (Maxine) **PLANK**: originated the quantum theory

Joseph (Josephine) **PRIESTLEY**: discovered oxygen

Walter (Wanda) **REED**: proved mosquitoes transmit yellow fever

Bertrand (Bernadine) **RUSSELL**: philosopher, founder of modern logic

Ernest (Ernestine) **RUTHERFORD**: discovered the atomic nucleus

James (Jane) **WATT**: invented the steam engine

Norbert (Nora) **WIENER**: founder of the science of cybernetics

Ferdinand (Fern) **ZEPPELIN**: airship designer.

A / COMPARE

In this chapter: Ten containers of different sizes, each with a different symbol. The moon underfills the sun, but overfills the star. The moon equals 4 hearts! Is this science or is this poetry? The moon and heart have equal height but unequal diameter. Let's compare top views and side views. Is this geometry or is this art? Rename the heart in terms of "x" and the moon becomes "4x." This sounds more like algebra. It doesn't really matter what you call it, as long as students are pouring lentils, comparing volumes and *learning*.

Basic Materials: Quantities define maximums needed to support any one Job Card in this chapter. Store *high-quantity basics* (Job Boxes, liters of lentils, bottle lids, scoops, funnels) on and under a table or counter. Store *low-quantity basics* near the "basics" sign (see page 45) or in a "basics" box. Consult our Glossary on pages 6-9 for a full description of these items. See the next page for additional special materials used in this chapter.

- [ ] **1 job box**
- [ ] **up to 2 liters of lentils**
- [ ] **1 scoop**
- [ ] **1 funnel**
- [ ] **masking tape**
- [ ] **tub**

☞ *Please observe our copyright restrictions on page 2.*

Store these chapter-specific items together in a designated place. They require about 1/2 square foot of dedicated space. General classroom materials (like scissors and tape) are also listed below when used, while others (like pencil and paper) are always assumed.

set of 10 containers: 4 glass jars and 6 plastic vials (compare 1, 2, 3, 4, 5, 6, 7)

Apply symbol labels, as detailed below, on each container in this set. Nest the smaller containers inside larger ones. Store them all together in a **gallon storage jug**.

Paper labels are provided for 3 identical sets. You will need this many to serve a class of 30 students working simultaneously in all chapters of this curriculum.

GLASS JARS

•CIRCLE: An **8 fl oz mayonnaise jar** or similar product. We used the smallest Best Foods brand mayonnaise jar available. Other brands may vary slightly in volume, but probably not enough to affect quantitative outcomes.

•TRIANGLE: a **6 oz baby food jar** (BFJ).

•MOON: a **4 oz baby food jar**.

•CLOUD: a **2¹/₂ oz baby food jar**. Look for the newer Gerber brand wide-mouth version, redesigned so the same lid fits all sizes. Its capacity is slightly larger than the older small-mouth version.

PLASTIC VIALS

•SQUARE: A **T–60 dram I-O plastic vial**, also called a standard cup. Do NOT substitute I-O 60 dram vials without child-proofing nubs around the mouth. These hold significantly less.

•SUN: A **T–40 dram I-O plastic vial**.

•STAR: A **T–30 dram I-O plastic vial**.

•FISH: A **T–16 dram I-O plastic vial**.

•PERSON: A **T–13 dram I-O plastic vial**. Some pharmacies do not to carry this size because it is so similar to a T-16. Ask at other stores, or purchase from TOPS. Alternately, cut to size a T-16 vial to 65 mm, the height of an equal-diameter T-13.

•HEART: A **T–8¹/₂ dram I-O plastic vial**.

4 puzzle books (compare 3, 4, 5)

Make single photocopies of the 4 line masters. Assemble as directed under booklets.

equation tags (compare 3, 4, 5, 6, 7)

Photocopy the line master on page 24 and assemble as directed in the grey section. Store these tags in a **clear cup**.

Masking tape label: *equation tags*.

set of 10 containers:
4 glass jars and 6 plastic vials
(compare 1, 2, 3, 4, 5, 6, 7)

4 puzzle books
(compare 3, 4, 5)

equation tags
(compare 3, 4, 5, 6, 7)

A / COMPARE
special materials

COMPARE
4 jars
(circle, triangle, moon, cloud)
6 vials
(square, sun, star, fish, person, heart)

COMPARE
4 jars
(circle, triangle, moon, cloud)
6 vials
(square, sun, star, fish, person, heart)

COMPARE
4 jars
(circle, triangle, moon, cloud)
6 vials
(square, sun, star, fish, person, heart)

CHAPTER SIGN: Glue to a 4 x 6 inch index card, and fold in half. Stand this sign in the space where you store special materials for this chapter. Or cut this sign in half, and glue both pieces to a grocery bag that has been cut to size, or to a box.

Apply each label to a gallon storage jug with clear packaging tape. (Three duplicate sets recommended for large class sizes.)

Apply each label to the container specified under special materials. Use clear packaging tape. Make multiple copies for multiple sets.

circle	square	triangle	sun	moon

  |  |  |

star	cloud	fish	person	heart

TEACHING NOTES

Purpose

To experimentally compare volumes by pouring lentils from one container into another. To order containers by size from largest to smallest.

Introduction

♦ Demonstrate how to fill a container *fair and full*. This is a simple 2-step process:

(1) *Overfill* with lentils, leaving a mound on top.

(2) *Shake once* gently to remove excess. (The top remains slightly rounded.)

♦ It is also important to recognize techniques that result in measures that are *not* fair and full:

• *Do not* overfill and pat down. This compresses the lentils, resulting in *more* than a standard measure inside.

• *Do not* overfill and scrape level. This drags out the top layer of lentils, leaving *less* than a standard measure inside.

♦ Pour a liter of lentils into a job box, and arrange the 4 glass jars inside, from largest to smallest.

• Write "large overfills small" on your blackboard. Demonstrate this principal by "pouring down the line." Fill the circle jar fair and full. Notice how it overfills the triangle; how the triangle overfills the moon; how the moon overfills the cloud.

• Write "small underfills large" on your blackboard. Demonstrate this principal by "pouring up the line." Fill the cloud fair and full. Notice how it underfills the moon; how the moon underfills the triangle; how the triangle underfills the circle. (Add additional lentils after each pouring to bring underfilled containers up to fair and full.)

Focus

◆ This circle jar looks like it holds more lentils than this triangle. Can you prove this by pouring? (Fill one fair and full and pour it into the other. Beginners may fill *both* containers, then do a lot of arm waving about which appears to hold more.)

◆ Can you line up all 10 containers from largest to smallest?

Checkpoint

◆ Please fill the _____ fair and full.

◆ I see that the square is taller than the circle. Does this mean the square holds more? Show me.

◆ Containers arranged from largest to smallest: 1) circle, 2) square, 3) triangle, 4) sun, 5) moon, 6) star, 7) cloud, 8) fish, 9) person, 10) heart.

◆ Rule: Large overfills small, and small underfills large.

A/1 You need...

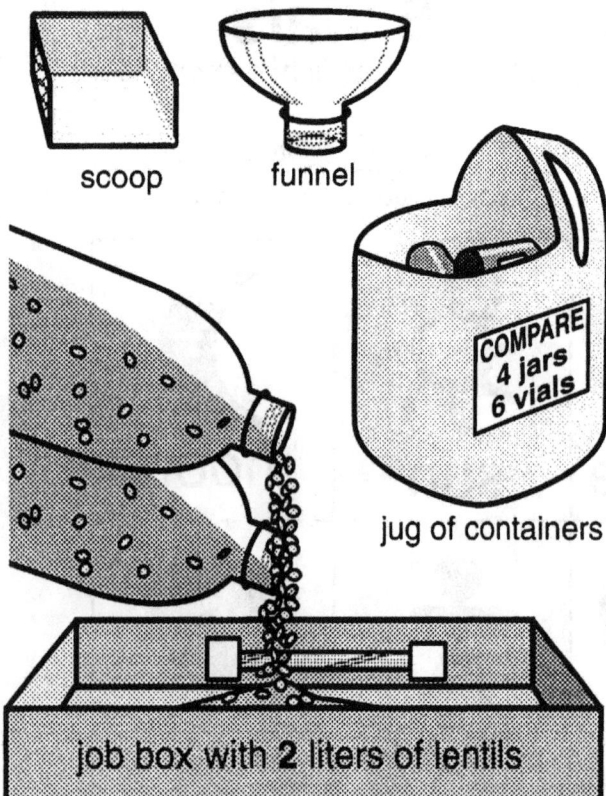

scoop   funnel

jug of containers

COMPARE
4 jars
6 vials

job box with **2** liters of lentils

compare 1

Line up all 10 containers in order of size. List them with the largest as #1, and work down to smallest.

Which holds more?

Fill #1 with lentils. Pour in your list order:

#1   #2   #3   #4 ....#10

Did you put any containers in the wrong order? Change your list and test again.

Write a rule using these words:

✓*large* ✓*small* ✓*overfills* ✓*underfills*

TEACHING NOTES

Purpose
To draw containers, both top and side views, to actual size. To compare diameters and heights.

Introduction
▶ Hold up a tin can. Observe it and draw it actual size from these perspectives:

• Top View: Looking from above, we see a circle. (Hold the top to your blackboard and trace the circumference.) Notice I begin by reaching way around the can, so I can draw the full circle without lifting my hand. Notice, too, that I can hold the chalk at any angle and still draw a perfect circle.

• Side View: Looking at the can from the side, we see a rectangle. (Hold the side to your blackboard and trace its perimeter.) Notice how the chalk touches the can in different places as I draw. I must hold the chalk perpendicular to the drawing surface, on all sides, to trace an actual-sized rectangle with good proportions.

▶ Define and discuss these lengths:

• Height: How tall the container is.

• Diameter: How wide the container is across the center of its top or bottom circle.

Focus
◆ Let me see you draw a top view and a side view of the sun container. Be sure to label the height and diameter, as you see in the Job Card.

◆ If you trace the mouth of the circle jar, will this top view be full size? *No, it will be a little less than full size because the body of the jar is wider.*

Checkpoint
May I see your drawings? Did you label each one and indicate both height and diameter?

More
◆ Find a more accurate way to draw a side view of a vial. Here's one method: Tick marks at top and bottom define height when the container lays on its side. Tick marks at opposite sides of the mouth define diameter when it stands on end. Complete the drawing with a straightedge.

◆ Make a contour gauge to "read" the shape of a jar: Cut a strip of cardboard longer than the jar, so the corrugated ridges run crosswise. Cover a long edge with 1 or 2 layers of masking tape, and push a round toothpick halfway into each corrugation, through the tape. Push the points against the jar so the toothpicks copy the shape, then lay the gauge on paper, and draw along the points. Find the jar's width with this gauge, too!

A/2 You need...

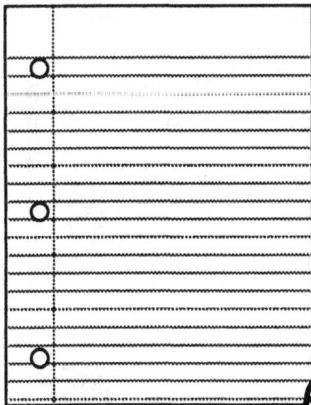

notebook paper

jug of containers

---

compare 2 👤 or 👤👤

Draw a TOP VIEW and a SIDE VIEW of the sun, fish and heart containers. Label your drawings.

TOP VIEW:

SIDE VIEW:

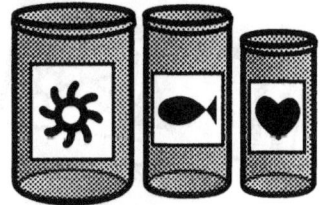

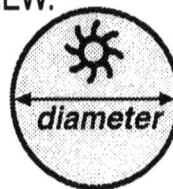

diameter

height

Draw the circle and moon in the same manner. Label both height and diameter.

TEACHING NOTES

Purpose

To compare volumes by pouring lentils. To express inequalities as written equations.

Introduction

♦ Review fair and full: Overfill and shake once. (This is an abbreviated way of saying: overfill, then gently shake away the excess lentils piled on top.)

♦ Demonstrate that lentils slightly compress (squeeze together) when shaken down.

• Fill the circle jar fair and full. Notice that the top is very slightly rounded.

• Place your palm over the top, slightly cupped to provide a little space for shaking. Vigorously shake the lentils up and down to settle them as much as possible. Notice that the same quantity of lentils no longer quite reaches to the rim of the jar. The contents were settled by shaking.

• Implications: Fair and full is always based on loose fill. Avoid settling by excessive shaking.

♦ Introduce the Puzzle Book:

• Discuss the meaning of the inequality symbols. The *small* pointed side of the symbol is always on the side of *less*. The *large* open side of the symbol is always on the side of *more*. (Or, hungry alligators always go after the biggest meal.)

• Notice that each leaf of the book is folded into a double layer: don't try to separate them. The puzzle is on the front of each page and the answer is on the back. Hold the book open by sharply creasing back the pages. Then prop it, open-faced, in the lentils.

• Do the first problem as an example. Turn the page to confirm you answered correctly.

♦ Use Equation Tags to arrange the inequality shown on the Job Card. Pour lentils to prove that it is true. (A sun and 3 hearts don't quite fill the circle.)

Focus

◆ Have you solved all the Puzzle Book problems? Let me see you try this one.

◆ Can you arrange and write your own inequalities using Equation Tags?

Checkpoint

◆ Did you solve all 8 puzzles? Show me how to solve this one.

◆ May I see your written work? Show me that this inequality you have recorded is correct.

A/3 You need...

puzzle book

equation tags

COMPARE 4 jars 6 vials

job box with **2 liters** of lentils

---

compare 3 🚹

Solve each puzzle:
Choose the correct inequality symbol.

compare 3 puzzle book

choose

Set up, test, and write your own inequalities.

more > less

☀ + 3

● > ☀ + 3 ♥

1 | **1**

choose

more > less

2 | **2**

choose

less < more

3 | **3**

choose

less < more

4 | **4**

choose

more > less

5 | **5**

choose

less < more

6 | **6**

choose

more > less

7 | **7**

choose

more > less

8 | **8**

choose

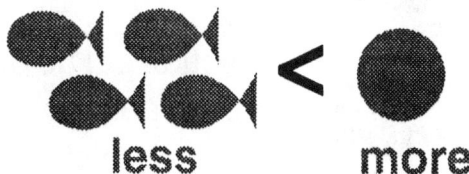
less < more

Preparation: 1 copy

DIRECTIONS:

1. Cut this page into 2 rectangles along the dashed lines.
2. Fold each rectangle in half, with numbers on the front and grey on the back.
3. Fold each in half again lengthwise to hide the grey inside.
4. Tape cut edges together along the entire length of each folded strip to make 2 very long stand-up signs.
5. Cut apart along the solid lines to make 12 stand-up signs.
6. Glue front and back layers together on each sign.

EQUATION TAG:

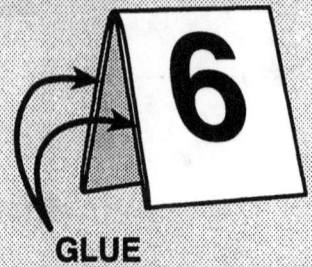

GLUE

same

=

=

same

2

2

3

3

4

4

5

5

6

6

more

>

>

more

less

less

7

7

8

8

+

+

−

−

÷

÷

EQUATION TAG:

GLUE

DIRECTIONS:

1. Cut this page into 2 rectangles along the dashed lines.
2. Fold each rectangle in half, with numbers on the front and grey on the back.
3. Fold each in half again lengthwise to hide the grey inside.
4. Tape cut edges together along the entire length of each folded strip to make 2 very long stand-up signs.
5. Cut apart along the solid lines to make 12 stand-up signs.
6. Glue front and back layers together on each sign.

Preparation: 1 copy

TEACHING NOTES

Purpose

To decide by experiment if various combinations of containers hold more, less or equal volumes. To discover and record equalities involving addition and subtraction.

Introduction

▶ Try the first problem in the Puzzle Book as an example. Fill the fish and person containers fair and full. Discuss all possible outcomes *before* pouring them into the empty star.
- If the star is *underfilled*, it must hold *more*.
- If the star is *overfilled*, it must hold *less*.
- If the star is *filled fair and full*, it must hold *an equal amount*.

▶ Show schematically how addition and subtraction require different pouring operations:

- *Addition* means to combine quantities.

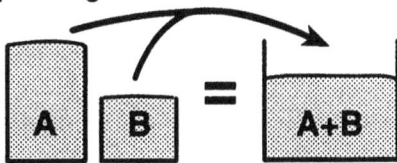

- *Subtraction* means to take away and see what's left.

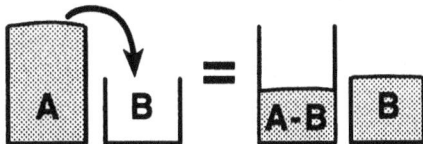

▶ When pouring from a large container into a smaller one, you need a funnel to keep from spilling. Or use your hand like a funnel.

Focus

◆ Have you solved all the Puzzle Book problems? Let me see you try this one.

◆ Can you arrange and write your own equalities using Equation Tags?

◆ Can you arrange an equality between ____ and ____? (See below for possible pairs.)

Checkpoint

◆ Did you solve all 8 puzzles? Show me how to solve this one.

◆ May I see your written work? (Inequalities are abundant; equalities are harder to find. Here are some involving addition and subtraction not already used in the Puzzle Books.)

$$\text{heart} + \text{fish} = \text{cloud}$$
$$\text{triangle} - \text{heart} = \text{sun}$$
$$\text{heart} + \text{square} = \text{circle}$$
$$\text{square} - \text{heart} = \text{triangle}$$
$$\text{cloud} + \text{moon} = \text{square}$$
$$\text{star} - \text{person} = \text{fish}$$

A/4 You need...

puzzle book

equation tags

COMPARE 4 jars 6 vials

job box with **2** liters of lentils

---

compare 4

Solve each puzzle:
Choose the correct sign.

compare 4 puzzle book — choose

Find your own equalitites and inequalities. Write them down.

1 | 1

choose
▧◁ + 👤 ⋁⋁⋁ ★

same ▤ same

2 | 2

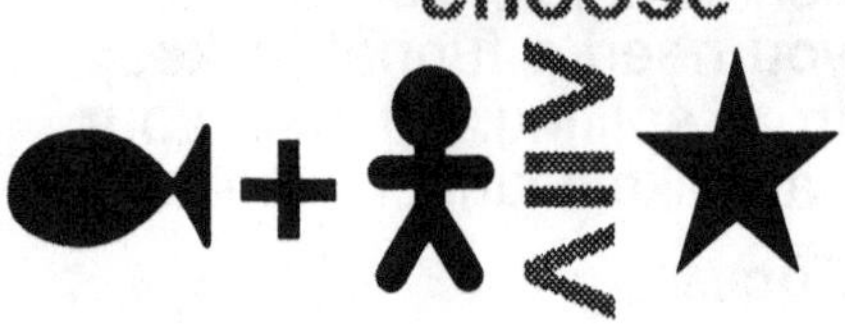

choose
♥ + ☁ ⋁⋁⋁ 🌙

less < more

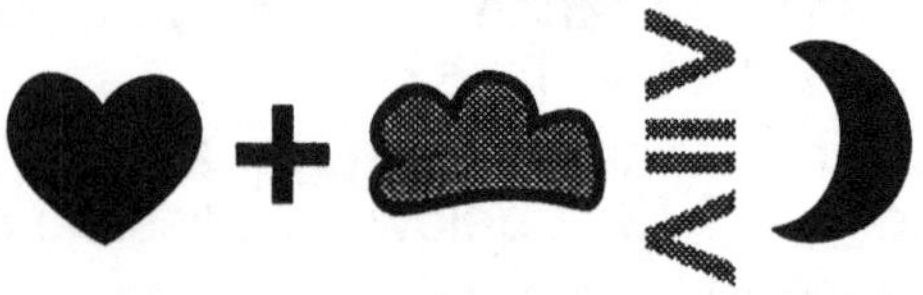

3 | 3

choose
♥ + ★ + ☁ ⋁⋁⋁ ■

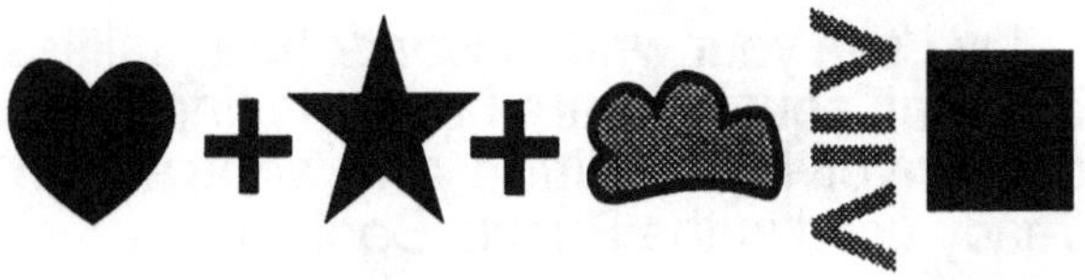

same ▤ same

4 | 4

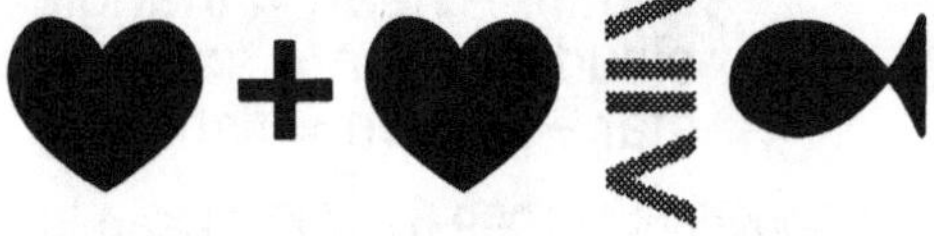

choose
♥ + ♥ ⋁⋁⋁ ◁◗

more > less

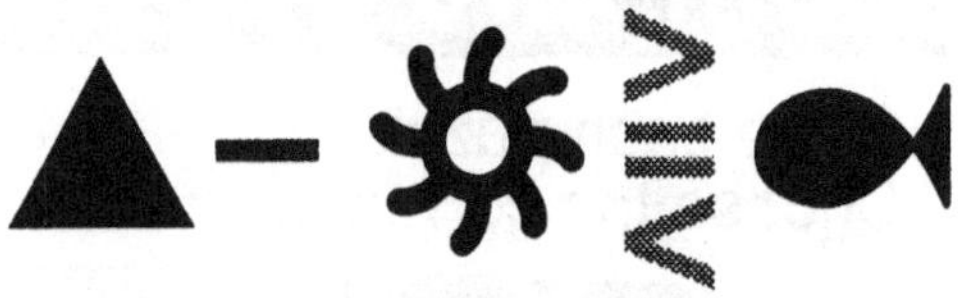

5 | 5

choose
▲ − ✹ ⋁⋁⋁ ◗◁

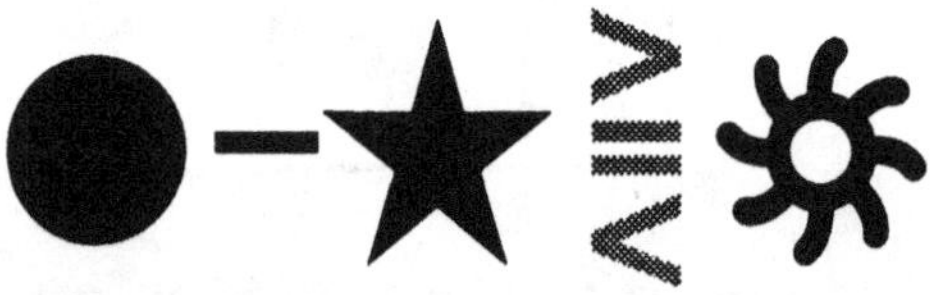

less < more

6 | 6

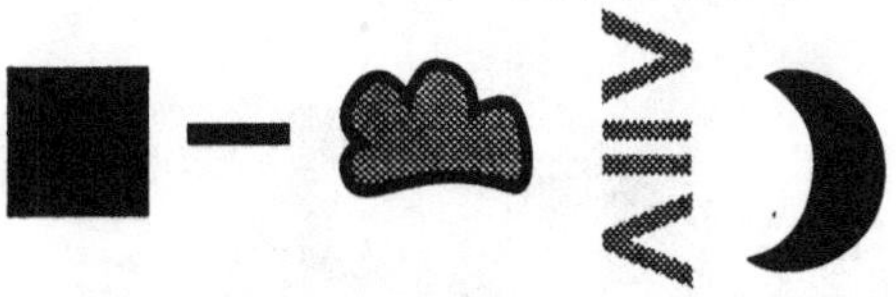

choose
● − ★ ⋁⋁⋁ ✹

same ▤ same

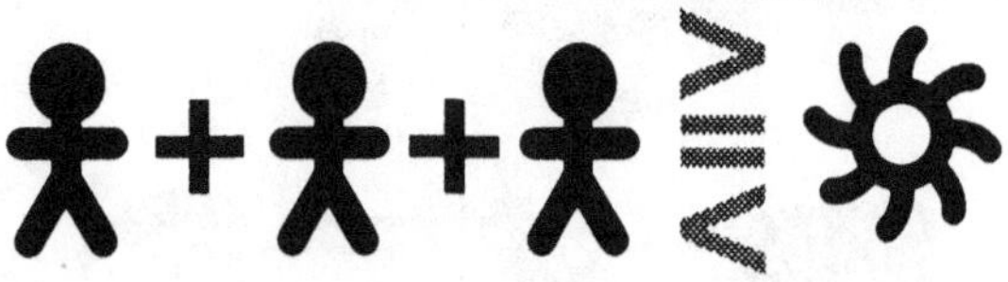

7 | 7

choose
■ − ☁ ⋁⋁⋁ 🌙

same ▤ same

8 | 8

choose
👤 + 👤 + 👤 ⋁⋁⋁ ✹

more > less

Preparation: 1 copy

TEACHING NOTES

Purpose

To understand addition, subtraction, multiplication and division in terms of pouring lentils. To discover and record equalities.

Introduction

Demonstrate 4 basic pouring operations with the sun cup and person cup. Always pour the final result into 1 or more tubs:

- **sun + person**: Fill both fair and full. Pour them into a tub.
- **sun – person**: Fill the sun fair and full. Pour this sun into the empty person until the person is fair and full. Pour the remainder into a tub.
- **3 suns**: Fill the sun fair and full 3 times. Pour each addition into a tub.
- **sun/3**: Fill the sun fair and full. Divide it equally between 3 tubs.

Focus

◆ Have you solved all the Puzzle Book problems? Let me see you try this one.

◆ Can you invent your own puzzle problems for others to solve?

Checkpoint

◆ Did you solve all 8 puzzles? Show me how to solve this one.

◆ May I see your written work? Show me that this equality/inequality you have recorded is true.

◆ More equalities not in the Puzzle Books:

person + 2 clouds = square
circle – 2 hearts = triangle
heart + person + 2 fish = triangle
person + 3 fish = square
heart + person + fish + cloud = square

More

◆ Rearrange Puzzle Book equations to make new ones. Test them by pouring lentils:

moon + fish = triangle (so) moon = triangle - fish
4 persons = triange (so) person = triangle/4

◆ Look for equalities that result from multiplication on both sides. Use a tub to collect all the lentils on one side of the equation before transferring them to the other side.

3 clouds ⤵ = 2 moons

◆ Collect your own small containers: creamer cups, deli condiment cups, film cans, plastic vitamin bottles, etc. Cut to size, as necessary, so each holds a different volume. Label them with symbols or letters of your choice. Explore relationships and develop equations. You might even invent a puzzle book for classmates to solve.

A/5 You need...

puzzle books

equation tags

COMPARE 4 jars 6 vials

job box with 2 liters of lentils

compare 5 👤

Solve each puzzle.

compare 5 puzzle book A

🌙 + 🐟 = ?

compare 5 puzzle book B

☁ + 👤 = ?

Make up new puzzles on folded squares of paper, with answers inside. Test them on each other.

⬛ – ❤ = ?

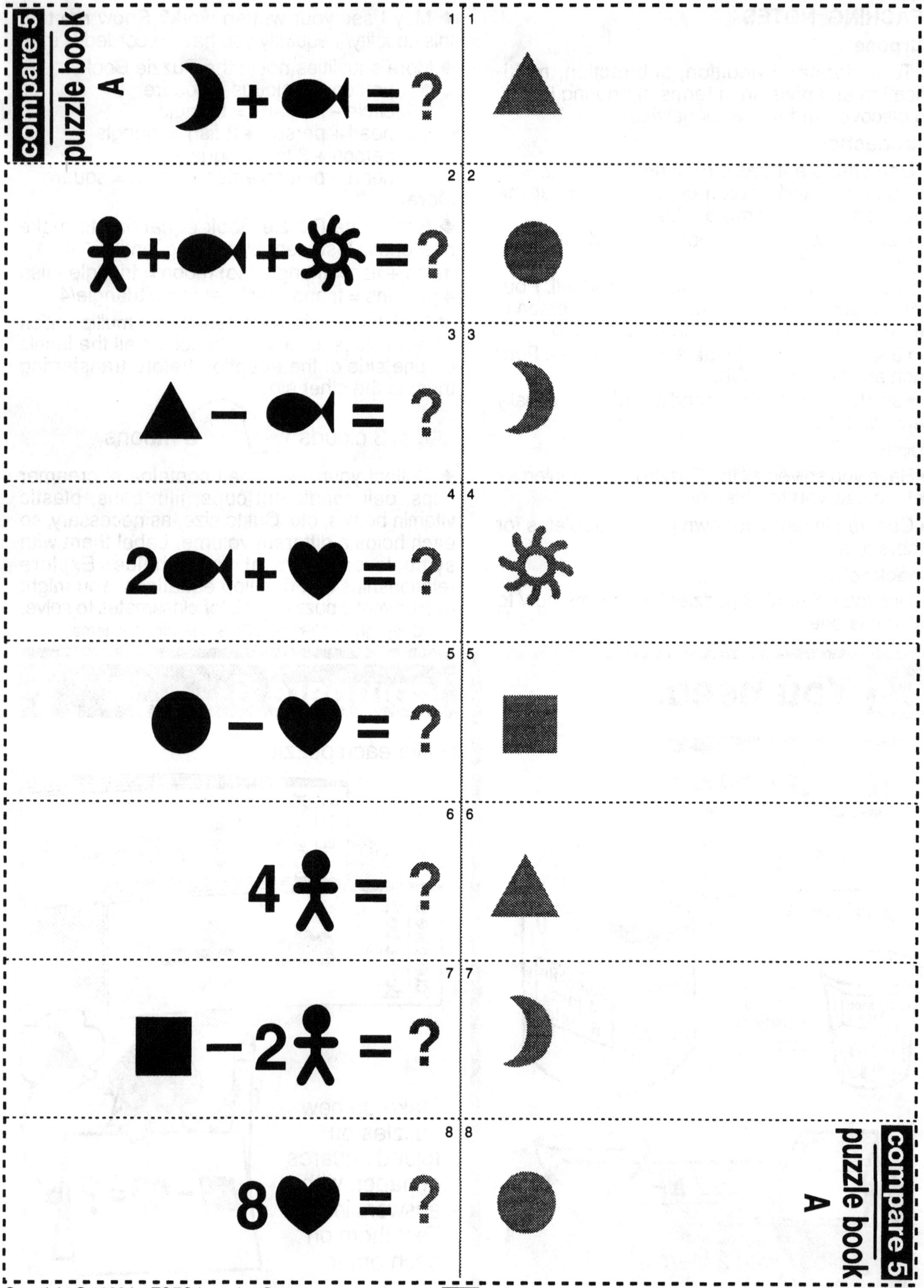

1 🌙 + 🐟 = ?

1 ▲

2 🧍 + 🐟 + ☀ = ?

2 ⬤

3 ▲ − 🐟 = ?

3 🌙

4 2🐟 + ♥ = ?

4 ☀

5 ⬤ − ♥ = ?

5 ⬛

6 4🧍 = ?

6 ▲

7 ⬛ − 2🧍 = ?

7 🌙

8 8♥ = ?

8 ⬤

Preparation: 1 copy

1 | 1

�️ + 🧍 = ?   🌙

2  | 2

2 🐟 + 🧍 + ❤️ = ?   ▲

3 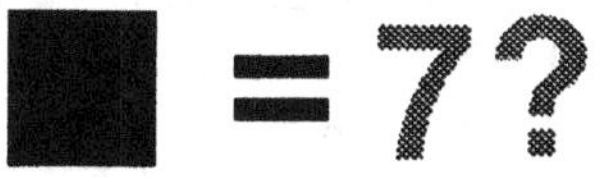 | 3

⬛ = 7 ?   ❤️

4 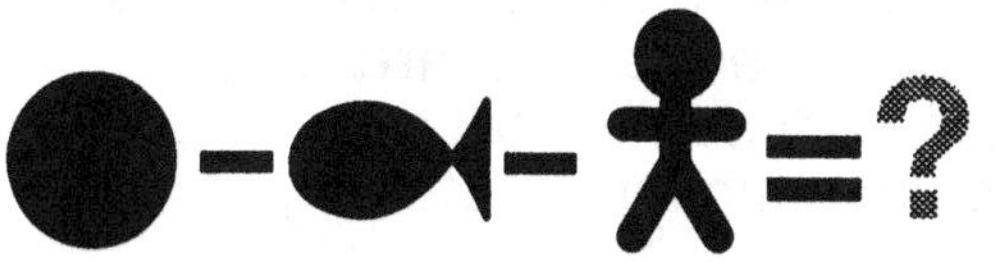 | 4

⬤ – 🐟 – 🧍 = ?   ☀️

5 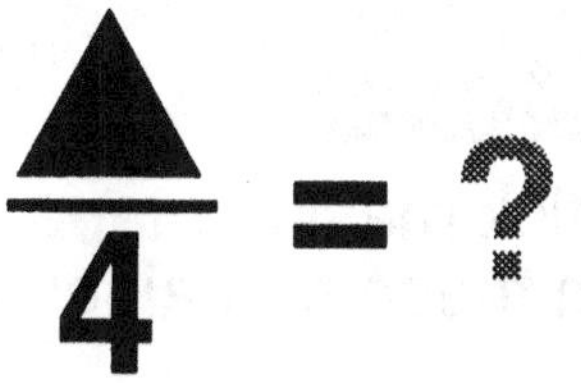 | 5

$$\frac{▲}{4} = ?$$ 🧍

6 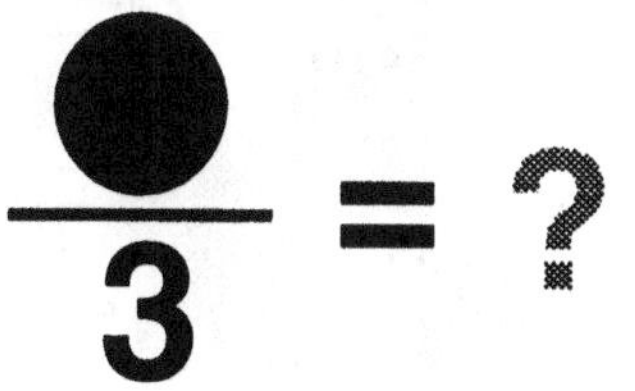 | 6

$$\frac{⬤}{3} = ?$$ ☁️

7 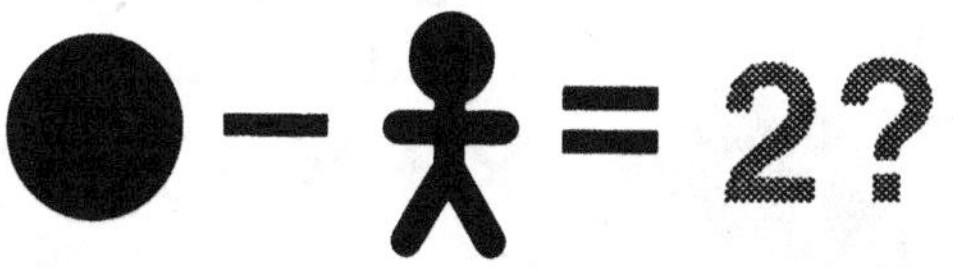 | 7

⬤ – 🧍 = 2 ?   ★

8  | 8

⬛ – 🧍 = 2 ?   ☁️

Preparation: 1 copy

TEACHING NOTES

Purpose

To explore relationships between 4 containers that have been redefined in terms of a unit volume. To discover and record simple algebraic sums.

Introduction

◗ Look at this container labeled with a heart. In this Job Card we are going to give it a second name, "x." I'll write this letter on a masking tape label I can remove when the lesson is over. (Apply the tape so both names are visible.)

◗ Let's name these other 3 containers all in terms of the **heart** container, called **x**.

• **moon**: 4 x's (pour in four hearts) fill the moon, so we'll call the moon **4x**. (Add label.)

• **triangle**: 6 x's (pour in six hearts) fill the triangle, so we'll call the triangle **6x**. (Add label.)

• **circle**: 8 x's (pour in eight hearts) fill the circle, so we'll call the circle **8x**. (Add label.)

◗ Let's write equations two ways, using both names:

$x + x + x + x = 4x$ *(or)* $4\heartsuit = \,$)

$x + x + 4x = 6x$ *(or)* $2\heartsuit + \,$) $= \triangle$

$4x + 4x = 8x$ *(or)* 2) $= \bullet$

(Note: Write 4 hearts as $x + x + x + x$, reserving the name **4x** for the moon. Similarly, write 2 moons as **4x + 4x**, reserving the name **8x** for the circle.)

Focus

◆ Have you labeled all 4 vials with new names?
$\heartsuit = x;$) $= 4x;$ $\triangle = 6x;$ $\bullet = 8x.$

◆ Can you write a relationship between heart and triangle using only x's? (Student should write $x + x + x + x + x + x = 6x$, not $6x = 6x$.)

Checkpoint

◆ Can we call the moon 5x instead of 4x? Why?

◆ May I see your written work? Show me that this pair of equations say the same thing.

$x + x + x + x = 4x$ *(or)* $4\heartsuit = \,$)

$x + x + x + x + x + x = 6x$ *(or)* $6\heartsuit = 1\triangle$

$x + x + 4x = 6x$ *(or)* $2\heartsuit + \,$) $= \triangle$

$x + x + x + x + x + x + x + x = 8x$ *(or)* $8\heartsuit = \bullet$

$x + x + x + x + 4x = 8x$ *(or)* $4\heartsuit + \,$) $= \bullet$

$x + x + 6x = 8x$ *(or)* $2\heartsuit + \triangle = \bullet$

$4x + 4x = 8x$ *(or)* 2) $= \bullet$

$4x + 6x = 8x + 2x$ *(or)*) $+ \triangle = \bullet + 2\heartsuit$

$4x + 4x + 4x = 6x + 6x$ *(or)* 3) $= 2\triangle$

$4x + 8x = 6x + 6x$ *(or)*) $+ \bullet = 2\triangle$

$4x + 6x + 6x = 8x + 8x$ *(or)*) $+ 2\triangle = 2\bullet$

$6x - x - x = 4x$ *(or)* $\triangle - 2\heartsuit = \,$)

$6x + 6x - 4x = 8x$ *(or)* $2\triangle - \,$) $= \bullet$ *...and more...*

(Some equations are only approximate. If necessary, say "almost equal." Some relationships require a tub for transferring.)

You need...

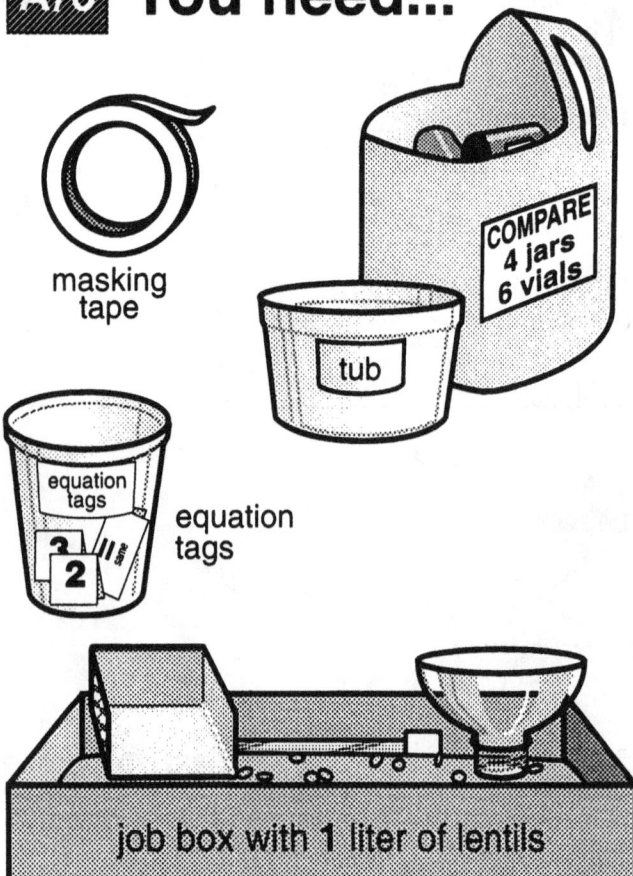

masking tape

tub

COMPARE 4 jars 6 vials

equation tags

equation tags

job box with **1** liter of lentils

compare 6

Stick masking tape labels on these containers:

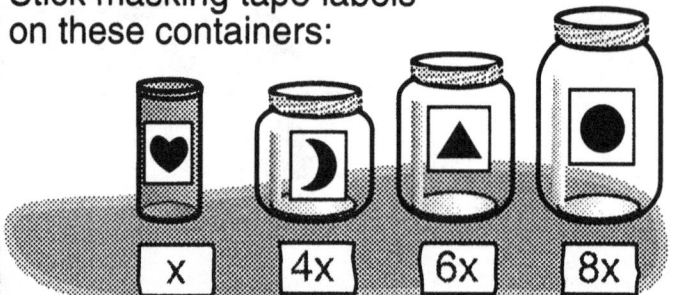

| x | 4x | 6x | 8x |

Write equations using x's first, and then symbols. For example:

Check.

$$X+X+X+X=4X$$
$$4\heartsuit = \,)$$

Peel off the new tape labels when you finish.

TEACHING NOTES

Purpose

To provide students with an approved way to pursue their own ideas about comparing volumes. To encourage creativity.

Introduction

Display this card whenever you want to do this suggested activity or experiments you design yourself.

Focus

◆ What do you want to study about comparing?

◆ Are these the materials you need?

Checkpoint

Report in words and pictures:
- What you did.
- What you learned.
- Questions you may still have.

Special Note

Use the opposite page to record especially clever ideas. Use them to inspire students that follow to be even more creative and inventive.

Send TOPS these ideas, too. We may include some in a future edition of this book, as sparks to ignite the imaginations of future young scientists.

Better yet, let your students address their own ideas to the world! Suggest that they write a report, and send it to us at the address on the title page. We'll send an affirming "thank you" note.

_____
_____
_____
_____
_____
_____
_____
_____
_____
_____
_____
_____
_____
_____
_____
_____
_____
_____

A/7 You may use...

Your own containers of different sizes, labeled with your own symbols.

COMPARE
4 jars
6 vials

equation tags

job box with **2 liters of lentils**

☞ Ask your teacher for other items. Tell why you need them.

compare 7 🧍 or 🧍🧍

On your own.

My idea:
Arrange the symbols to tell a story:

It was a dark and 🌙⭐

and Joe's ❤ was flip-flopping

like a 🐟

A career is born...

B / SEARCH

In this chapter: The world keeps turning. There is political controversy at Lentil Lake: Juan is worried because pinto fish populations are way down, but Terry wants to fish with nets! Conditions in Lentil Land continue to evolve: penny rabbits have survived the pressures of natural selection; red rabbits are nearly extinct, but strange new adaptations with cunning camouflage emerge on a daily basis. Elsewhere, Tov and Ali are graphing seed distributions. Chris and Lon are perfecting new iron extraction technologies. Sue and Tom are mapping buried treasure.

Basic Materials: Quantities define maximums needed to support any one Job Card in this chapter. Store *high-quantity basics* (Job Boxes, liters of lentils, bottle lids, scoops, funnels) on and under a table or counter. Store *low-quantity basics* near the "basics" sign (see page 45) or in a "basics" box. Consult our Glossary on pages 6-9 for a full description of these items. See the next page for additional special materials used in this chapter.

- [ ] **1 job box**
- [ ] **1 liter of lentils**
- [ ] **scoop**
- [ ] **3 tubs**

- [ ] **1 bottle cap spoon**
- [ ] **1 clear cup**
- [ ] **10 craft sticks**

Store these chapter-specific items together in a designated place. They require about 1 square foot of dedicated space. General classroom materials (like scissors and tape) are also listed below when used, while others (like pencil and paper) are always assumed.

~75 pintos (search 1, 2)

Buy a **bag of pinto beans** from your grocery store, or sort them out of the bean mix also used in this chapter. Fill a **film can** to capacity and snap on the lid.

Paper label(s) provided on next page. Use extra to serve additional students.

screen (search 1, 2, 3, 4, 5, 8)

Purchase ¼ **inch grid hardware cloth** (wire screen) from a hardware store. Other mesh sizes will NOT work. Use **wire cutters** to cut a full 6 x 6 inch piece *plus* an extra fringe of half squares all around the perimeter. Cover exposed edges with ³/₄ inch **plastic electrical tape**, folded evenly over the wire and stuck to itself. Trim the tape at each corner. Bend up a half inch lip all around to create a shallow, square dish.

paper plates (search 1, 2, 3, 4, 8)

Use a generic, 9-inch diameter picnic plate, or equivalent.

10 pennies, 10 pintos, 10 reds (search 3)

Sort red beans and pintos from the **bean mix** used in this chapter. Place these seeds and the pennies in the same **film can**, and snap on the lid.

Paper label provided.

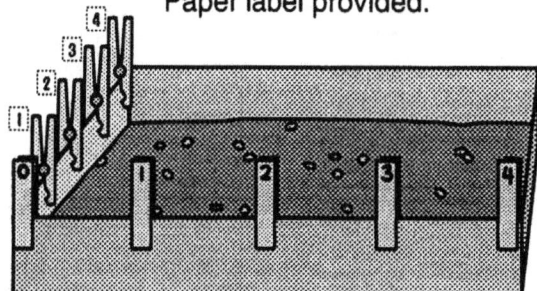

bean mix (search 4, 8)

Purchase two 1-pound bags of **15- or 16-variety soup mix**. Try to find a mix that contains dried whole peas, a very useful seed not available in all mixes. (Try Western Family brand in western states.) Fill a **quart jar** about ³/₄ full, and screw on the **lid**.

Masking tape label: *bean mix*.

paper clips (search 6, 7)

Just a few.

compass lid (search 7)

Use a small **white bottle lid** (from a liter or 2 liter bottle) with no printing on the top. Label it like a compass: N, S, E and W.

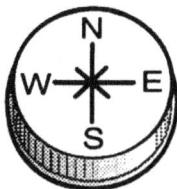

hand lens (search 8)

Any kind is OK.

4 magnets (search 5, 6, 7, 8)

Purchase rectangular, **ceramic refrigerator magnets** from an electronics store, science supply catalog, or directly from TOPS. Ours measure 1 inch x ³/₄ inch x ¹/₈ inch.

Wrap a strip of masking tape around three of these magnets, and label the poles N for north-facing, S for south-facing. Stand a new **brad** on its head at the center of the north pole on the fourth magnet. Wrap with masking tape to hold it in place. Attach a short length of **string** with more tape, and label the poles. Any number of magnets can now hang from the string with poles oriented up and down. Interesting pendulum interactions are possible, because the brad keeps repelling magnets from easily flipping over.

wood blocks (search 5)

Cut two **2 x 4's** to about 15 inches, the width of your Job Box.

30 BB's, 1 bottle cap (search 5)

Purchase BB shot from the camera and ammunition case of your local drug or variety store. Place in a **film can**, along with an unbent twist-off **bottle cap**, and snap on the lid.

Paper label provided.

medium jar lid (search 5)

About the size you find on a quart of mayonnaise.

grid markers (search 6)

Number **clothespins** from 0-4 and from 1-4 (labeling 9 altogether) with a fine-tipped **permanent marker**. Write directly on the clothespin wings, or use masking tape tags. Store these in a **clear cup**.

Paper label provided.

~75 pintos (search 1, 2)
screen (search 1, 2, 3, 4, 5, 8)
paper plates (search 1, 2, 3, 4, 8)
10 pennies, 10 pintos, 10 reds (search 3)
bean mix (search 4, 8)
4 magnets (search 5, 6, 7, 8)
wood blocks (search 5)
30 BB's, 1 bottle cap (search 5)
medium jar lid (search 5)
grid markers (search 6)
paper clips (search 6, 7)
compass lid (search 7)
hand lens (search 8)

B / SEARCH

special materials

CHAPTER SIGN: Glue to a 4 x 6 inch index card, and fold in half. Stand this sign in the space where you store special materials for this chapter. Or cut this sign in half, and glue both pieces to a grocery bag that has been cut to size, or to a box.

~ 75 pinto bean "fish"

~ 75 pinto bean "fish"

30 "rabbits"
10 reds
10 pennies
10 pintos

30 BB's 1 bottle cap

Apply each label to a film canister with clear packaging tape.

grid markers

Apply to clear cup with clear packaging tape.

Preparation: 1 copy

TEACHING NOTES

Purpose

To catch "pinto fish" by hand and by net over ten second intervals. To record the characteristic decline of a nonrenewed resource.

Introduction

▶ Show your class how to count a 10-second interval. Watch the second hand of a clock, while counting "1001, 1002, ..., 1010."

▶ Scatter all the pinto beans from a film can (70 to 75) over the lentils in the box and mix them in. Ask volunteers to "fish" the lentil "waters" in 10 second intervals. Record each catch in a blackboard table formatted like the tables below.

• What happened to the fish catch over time? *It diminished to almost nothing, because all the fish were caught.*

• Fish populations can bounce back if not driven to extinction. Can you think of resources that take a longer time to recover? *On a human time scale: oil, minerals, old growth trees, clean air, clean water, wilderness ...*

▶ How can you know you've caught all the fish?

• Count the beans in the film can before you start, and compare.

• Pack the found beans back into the film can and see how much room remains.

• Push all the lentils to one side of the box. Screen them systematically, shifting all filtered, pinto-free lentils on the other side of the box.

Focus

◆ Will you fish each season alone, or with a partner?

◆ How will you track your catch? *In a table.*

◆ Can you use "technology" to increase your catch? *Yes. Use a net!*

Checkpoint

◆ Table trends: The catch decreases slightly during the first 5 or 6 seasons, but remains remarkably stable at 8-12 fish per catch. Then the population crashes, in just a season or two, to 0-2 fish per catch.

◆ High technology has trade-offs. Yields are initially very high, but the resource is rapidly exhausted.

More

◆ Will 2 film cans of fish take twice as long to catch? *No.*

◆ What if you make the lake twice as deep, using 2 liters of lentils?

◆ Graph: seasons *vs* catch.

B/1 You need...

~75 pinto bean "fish"

screen

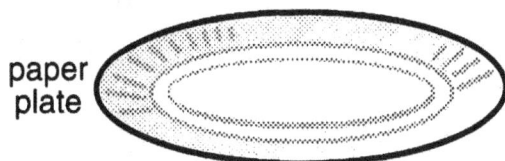

paper plate

scoop

job box with 1 liter of lentils

---

search 1   👤 or 👥

Nonrenewed Resources

Stock "Lentil Lake" with about 75 "pinto fish." Catch all you can, by hand, in 10-second "seasons." Tabulate your results.

me:

season	catch	total
#1		
#2		
#3		

you:

season	catch	total
#1		
#2		
#3		

Repeat this fishing experiment with a net.

What have you learned about nonrenewed resources? High technology?

TEACHING NOTES

Purpose

To understand the dynamics of a renewable resource. To maximize fish catches with wise management.

Introduction

♦ Define "carrying capacity," the maximum number of fish that can live in Lentil Lake and still have enough habitat and food to survive. In this activity it is 40 fish. (Ask if earth has a human carrying capacity. What might the number be? Are we close to reaching it?)

♦ Illustrate the basic natural laws in the Lentil Lake ecosystem by running these numbers through each blackboard table as a thought experiment:

NO FISHING:

season	fish population	my catch	your catch	survivors	offspring
#1	10	0	0	10	10
#2	20	0	0	20	20
#3	40	0	0	40	0
	40	0	0	40	0

NO LIMIT:

season	fish population	my catch	your catch	survivors	offspring
#1	10	10	0	0	0
#2	0	0	0	0	0
#3	0	0	0	0	0
	0	0	0	0	0

Is there compromise between these extremes? The future of Lentil Lake is in your hands.

Checkpoint

♦ Did you set limits to increase yields? Explain.
One possible solution is to limit fishing, or ban it altogether, until the lake reaches its maximum carrying capacity of 40 fish. Then enforce a 10-second "season" for 2 players, or a 20-second "season" for one.

♦ What have you learned? *Limits help more people catch more fish over the long term. People who don't like limits should be educated to see themselves as member of the human community. Limit laws should be enforced and violators fined.*

♦ Does this activity remind you of any resource management controversies recently in the news?

More

♦ The organization Net Fishing United is lobbying your office to allow its members to fish Lentil Lake with nets. A local environmental and sport fishing group called Lentil Lake Lookout is against any commercial fishing in this lake. Also, half of the lake, and the river leaving it, is on a reservation whose residents want to enforce their traditional rights to first catch. As head resource manager, you must help shape public policy. Be ready to defend your decisions and take the political heat.

♦ Graph: seasons *vs* catch.

B/2 You need...

screen

~75 pinto bean "fish"

scoop

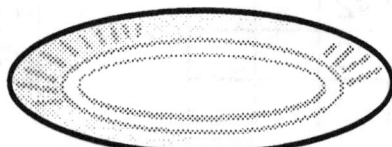

paper plate

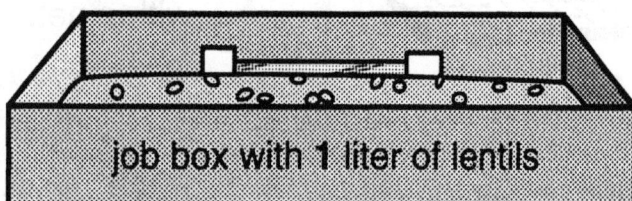

job box with 1 liter of lentils

---

search 2 👥

Renewable Resources

Will you set limits to manage the pinto fish in Lentil Lake? These are the laws of nature you must obey.

> *Starting Population:* 10 fish.
> *Carrying Capacity:* 40 fish.
> *Birth Rate:* Survivors double each year.

season	fish population	my catch	your catch	survivors	offspring
#1	**10**				
#2					
#3					

What have you learned about wise resource managment? What if some people don't like limits?

TEACHING NOTES

Purpose

To design 10 rabbits well suited to survive in a lentil habitat. To test your design against natural selection.

Introduction

▶ Mix all 30 "rabbits" from the film can into the lentils, then ask volunteer "foxes" to find them and put them in the paper plate "stomach." Call a halt to the search after about 25 rabbits all together have been "eaten."

▶ Which rabbit species best avoided predation? Discuss survival strategies. (Pintos hide better than Reds because of camouflage: they blend in better with the lentil background. Pennies hide best because they "burrow underground.")

▶ Invite students to play the rabbit game. Then challenge them to design their own new, improved species of rabbit that can survive even better. (All rabbits *must* be larger than the quarter inch screen.)

Focus

◆ Allow students to explore their own ideas before you step in with rabbit-building suggestions:

• Lentil clusters: Drop lentil-sized puddles of white glue into a small pile of lentils. Gently cover with more lentils. Wait a day, then dig out the "rabbits". Lightly rub off any loose lentils.

• Paper wads: Crumple up postage-stamp sized bits of a brown paper bag.

• Natural seeds: Screen a 15 or 16 variety soup mix to find seeds of suitable size. Whole peas work very well.

• Painted seeds: Paint in lentil browns and greens with a durable paint, such as acrylic.

• Found objects: Search for gravel with the appropriate shape, or bits of wood, or other debris.

• Clay: Sculpt realistic lentils, large enough to get trapped by the screen.

Checkpoint

◆ Students should first report on red rabbits, penny rabbits and pinto rabbits, giving numbers (data) to back up their claims.

◆ Students should next report on their own designer rabbits, again using hard data to compare survival rates.

◆ How does natural selection work? *Rabbits that survive the natural selector (the fox) produce offspring that look just like them. Rabbits that don't survive don't reproduce.*

More

Conduct a class contest. Declare a champion breed.

B/3 You need...

materials for designing rabbits:
glue, brown paper, bean mix,
paint, clay.

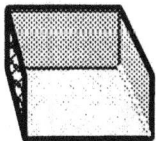

scoop

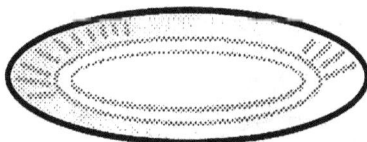

paper plate

screen –
use only for
final cleanup

30
"rabbits"
10 reds
10 pennies
10 pintos

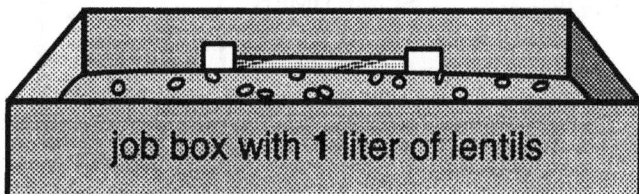

job box with 1 liter of lentils

search 3

Hide 30 "rabbits" in "Lentil Land." As a "hungry fox," you should hunt and capture all you can, up to about 25.

10 red rabbits   **10 penny** rabbits   **10 pinto** rabbits

Which rabbits survived to reproduce offspring like their parents?

Can you "adapt" 10 rabbits of a different species that survive better than these? (They MUST be large enough to separate by screening for final clean-up.)

TEACHING NOTES

Purpose

To sort a mix of seeds by size using a screen.

Introduction

♦ Filter just one screenful of mixed beans onto a paper plate. Pour the large seeds remaining on the screen into one tub; the small seeds falling to the paper plate into another tub.

♦ Screen each of these tubs of seeds a second time. Any of the *large seeds that do pass* the screen this time, and any of the *small seeds that don't pass*, are borderline cases. Put these medium seeds that *sometimes* pass through the screen into a third tub.

♦ Display a job sheet graph on your wall. Read through the various categories of seeds, from large on the left to small on the right.

♦ Sample a random spoonful from *one* of your three tubs to pour onto the empty paper plate. (Decide with your class whether you will select from the large, medium or small catogory.)

♦ Count one or two types of seeds from this spoon sample. Darken a corresponding number of circles to begin your bar graph. Speculate how seed distribution might shift if you sampled from the other two tubs, or from the original unsorted mix.

♦ Caution students never to mix the jar of bean mix into a box of lentils. It takes too much work to separate them out again! (If this should happen, screen out the larger seeds, then require the offender to separate *most* lentils from the remaining small seeds by hand. Or it may be easier to purchase another bean mix.)

Checkpoint

Inspect students' graphs. Up to 4 categories, A, B, C, and D, are possible. Discuss reasons for shifts in the frequency distribution. (After the seeds have been poured back into the main jar, notice how they retain their sorted distribution, with abrupt changes at the interface between large and small. After noticing, shake the jar to thoroughly remix the seeds.)

More

◆ Invent a way to sort seeds by shape. (Shake a random mix down a shallow incline in the job box. Round seeds tend to roll to the bottom, while flat seeds remain behind. These may be graphed into sorted seed distributions.)

◆ What forces sort sand, pebbles, stones, and boulders in nature? Consider the action of gravity, wind and water. Write a report.

B/4 You need...

paper plate

bottle-cap spoon

3 tubs

screen

job sheet

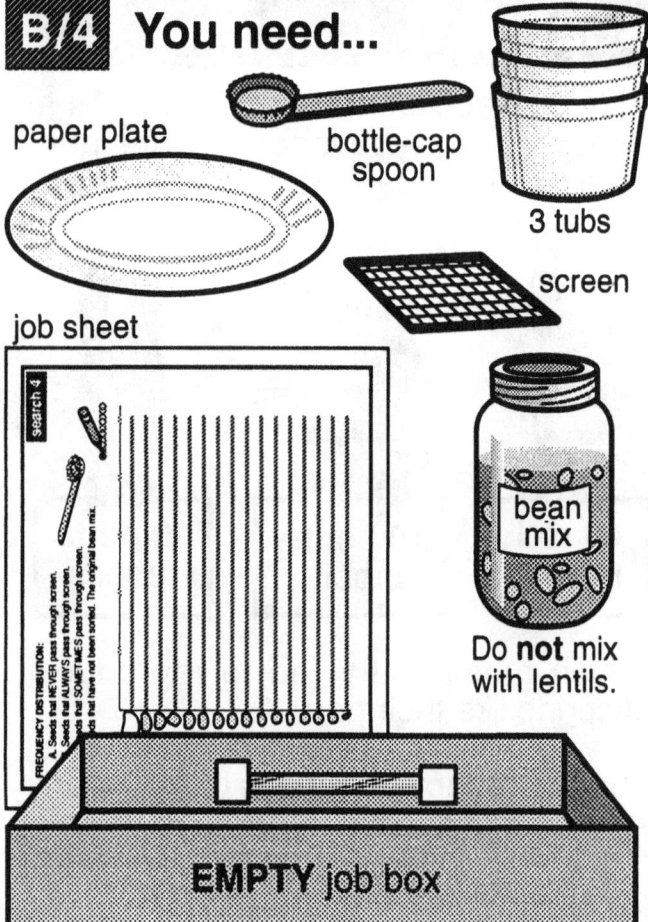

bean mix

Do **not** mix with lentils.

EMPTY job box

---

search 4

Sort a jar of bean mix. Rescreen the small and large seeds several times to develop a medium grade.

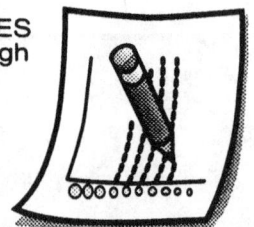

small seeds ALWAYS pass through

medium seeds SOMETIMES pass through

large seeds NEVER pass through

Graph a spoonful of each group of seeds.

Seed Distribution: Sample a spoonful and graph the numbers.

Circle the letter that defines this bar graph:

A. Seeds that NEVER pass through screen.

B. Seeds that ALWAYS pass through screen.

C. Seeds that SOMETIMES pass through screen.

D. Seeds that have not been sorted. The original bean mix.

	10	20	30	40	50
large lima					
red kidney					
pinto					
garbanzo					
baby lima					
pink bean					
medium white					
dark red					
black-eyed pea					
black bean					
whole pea					
small white					
green split pea					
yellow split pea					
lentil					
barley					

Job Sheet: several copies per student

TEACHING NOTES

Purpose

To search for BB's among the lentils in creative and inventive ways.

Introduction

▶ Pack BB's into a bottle cap as illustrated below and pass them around. They form a neat square of 4 in the middle, surrounded by an inner circle of 10, and an outer circle of 16. Pressing them into the bottle cap is a quick and easy way to know you have found all 30 BB's!

▶ *Pretend* to dump this capful of BB's into the lentils and mix them up. Challenge your class to invent innovative ways to find them again. Hold up magnets, the screen, and the scooper as interesting tools for exploration.

Checkpoint

Allow students to thoroughly explore their own ideas before you step in with suggestions:

• Using the magnets: Sweep systematically through the lentils; fish from the string; bulldoze a pair of magnets through the lentils by moving attracting magnets beneath the box (raise the box on 2x4 blocks).

• Using the scooper: Vigorously shake the scooper to spill the lightweight lentils over the raised edge, while the heavier BB's shift toward a lowered back corner; use a similar panning strategy with a pair of magnets stuck to the inside and outside of the scooper.

• Using a jar lid, clear cup, or screen: Press into the lentils from above, circling and pushing away lentils. These tools involve touch (BB's feel like ball bearings) and sight (a clear cup works like a glass-bottomed boat). A magnet in the cup allows you to release the BB's all at once. Hold the cup over the lid and lift the magnets out.

More

◆ Hold a class contest for extracting "gold" from Lentil Land. Challenge students to try methods and materials of their own, possibly in combination with suggestions above. Offer awards for the most efficient and most innovative ideas. Require students to submit drawings and written explanations of their plans.

◆ Invent a way to count BB's that doesn't use a bottle cap.

B/5 You need...

wood blocks

4 magnets

clear cup

30 BB's 1 bottle cap

screen

jar lid

scoop

job box with **1** liter of lentils

search 5   ♟ or ♟♟

Try different ways to separate BB's from lentils. List your discoveries.

Try:
✔ magnets;
✔ a scooper;
✔ a jar lid;
✔ a screen;
✔ a clear cup...

4
10
16
30 BB's

TEACHING NOTES
Purpose
To fish for paper clips buried in a coordinate grid. To practice plotting ordered pairs.

Introduction
▶ Set up a clothespin coordinate system as shown on the job card. Position the corner clothespins first (the 0 and 4's); then the middle clothespins (the 2's); then the remaining clothespins halfway between (the 1's and 3's).

▶ These numbers define an *x* axis (the front edge of the box) and a *y* axis (the left edge). Coordinates are pairs of numbers that help you locate a specific point that matches (or coordinates with) the *x* and *y* axes.

▶ Bury a paper clip:

• Starting from coordinate *(0,0)* in the left front corner of the job box, move right along the front of the box any distance *x*.

• Then move straight toward the back of the box any distance *y*.

• Place a paper clip on the lentils and record the coordinate *(x,y)* on your blackboard.

• Bury the clip. (This step is last because it is easy to forget where you buried it without first setting the coordinates.)

▶ Ask a volunteer to leave the room while you bury a second paper clip following the same procedure. Invite the volunteer back in to fish out both paper clips by lowering a magnet at the *exact locations* you have written on the board. If the coordinates are accurately recorded and interpreted, there is no need to hunt around.

Focus
◆ Can you place a paper clip at (__,__) on top of the lentils? Write this position down before you bury the clip.

◆ Turn around while I hide a paper clip. OK, I buried it at (__,__). Can you pick it out without hunting around?

◆ Can you bury a paper clip at ($1\frac{1}{2}$, $2\frac{1}{2}$)?

Checkpoint
◆ Show me your list of (x, y) coordinates. Let me see you bury another paper clip at the position you wrote here.

◆ Can you name the coordinates at each corner of the job box? *(0,0), (4,0), (4,4), (0,4).*

◆ What pair of coordinates marks the center of the box? *(2,2).*

◆ Which coordinates describe diagonal lines from corner to corner? *(0,0), (1,1), (2,2), (3,3), (4,4), and (0,4), (1,3), (2,2), (3,1), (4,0).*

B/6 You need...

paper clips

grid markers

4 magnets

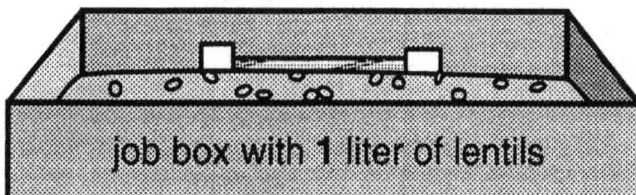

job box with 1 liter of lentils

search 6

Set up a clothespin grid on your Job Box.

Write a list of x, y coordinates. Bury a paper clip at each position you list.

Give your list to a friend to fish out the buried clips.

(3,4) is right here.

(x,y)
(3,4)

TEACHING NOTES

Purpose
To develop basic orienteering skills.

Introduction
▶ Direct students to letter a rectangular piece of scratch paper like a compass. Put N at the top; S at the bottom; E to the right; W to the left.

▶ Pretend the walls of your classroom run N-S and E-W, like the paper "compass." Post a large N near the center of one wall to define north. (If this is not actually north, tell your class this is an arbitrary assignment for this demonstration.)

▶ Ask students to orient their paper compasses to N as you've defined it. In a rectangular room, the sides of each paper should parallel the walls. North is a direction, not a location (except at the North Pole). Only those at the center of the room will see their "compass" aim directly at the "N" on the wall.

• Ask volunteers to identify the directions of objects and people that surround them. *The light switch is to my E; Kim is sitting NW of me; etc.*

• If two people sit facing each other, and they are both facing north, where are they? *They are sitting with the North Pole between them.*

• Direct a volunteer to any mystery object in your room by using compass directions.!

Focus
◆ This job card places N at the far side of the job box. Which side corresponds to S? E? W?

◆ Lay down these craft sticks following directions on the Job Card; following my directions.

Checkpoint
◆ Show me your written maps. Who tried to use them? Was he/she successful?

◆ Give me directions. I'll build a road.

More
◆ Can you build 2 roads to the same treasure?

◆ Play Ghost Writer: Think of a letter. Ask a friend to follow your directions to find out which letter you chose: *Lay 2 craft sticks end to end at the W side of the job box so they point N-S....*

◆ Trace around a deli tub lid to make a circle on paper.

• Box a compass: define N, S, E, W; then NE, SE, SW and NW on this paper circle.

• Cut it out and stick it to the lid with rolled tape.

• Go outside. If it is high noon, your shadow points due north. Otherwise, orient to a distant point on your N horizon.

• Draw a school yard map. Identify landmarks.

B/7 You need...

compass lid

4 magnets

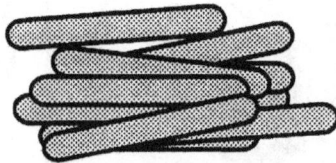

10 craft sticks

paper clip

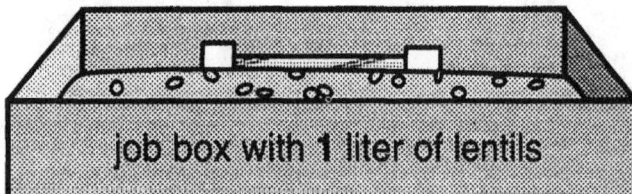

job box with 1 liter of lentils

---

search 7 ☥ or ☥☥

N

Bury a paper clip treasure.

Follow these directions with your compass. **X**

Begin at the **SE** *corner:*

> Go **W** 1 stick.
> Go **N** 1 stick.
> Go **NE** 1 stick.
> Go **N** 1 stick.
> Go **W** 2 sticks,
> Go **SW** 1 stick, and dig.

Pick up the sticks; turn around twice; lay down the sticks to find it again.

Write your own directions!
Bury a paper clip treasure so accurately that a friend can find it.

Begin at

TEACHING NOTES

Purpose

To provide students with an approved way to pursue their own ideas about searching through lentils.

Introduction

Display this card when you want to do the suggested activity or experiments that you design yourself.

Focus

◆ What do you want to study about searching?

◆ Are these the materials you need?

Checkpoint

Report in words and pictures:
- What you did.
- What you learned.
- Questions you may still have.

Special Note

Applaud creativity: Record some of your students' most amazing ideas on the opposite page.

B/8 You may need...

bean mix

Do **not** mix with lentils.

hand lens

EMPTY job box

☞ Ask your teacher for other items.
Tell why you need them.

search 8 — 🧍 or 🧍🧍

On your own.

Here's an odd one!

My idea:
Find and draw average specimens and weird specimens of different seed varieties...

C / MEASURE

In this chapter: Understand liquid measure without getting wet: third cups, half cups, cups, pints and quarts. We've even thrown in a tub that equals 3 cups. Pour and write 12 equations. Understand dry measure without getting lost: centimeters, inches and lentils. See and feel the difference between length, area and volume. Measure length with rulers and add up the units. Cover areas with grids and count the squares. Define volumes with paper boxes and count the unit volumes that stack inside.

Basic Materials: Quantities define maximums needed to support any one Job Card in this chapter. Store *high-quantity basics* (Job Boxes, liters of lentils, bottle lids, scoops, funnels) on and under a table or counter. Store *low-quantity basics* near the "basics" sign (see page 45) or in a "basics" box. Consult our Glossary on pages 6-9 for a full description of these items. See the next page for additional special materials used in this chapter.

☐ **1 job box**   ☐ **1 scoop**   ☐ **stick rulers**
☐ **up to 2 liters of lentils**   ☐ **1 funnel**   ☐ **clear tape**
☐ **extra cups**   ☐ **masking tape**   ☐ **bottle cap spoon**

☞ *Please observe our copyright restrictions on page 2.*

Store these chapter-specific items together in a designated place. They require about 1/2 square foot of dedicated space. General classroom materials (like scissors and tape) are also listed below when used, while others (like pencil and paper) are always assumed.

set of 6 cup measures (measure 2, 3, 8)

Nest these 6 containers inside each other and store them in a **gallon storage jug**. Two paper labels are provided. You may need an extra set to serve many students.

ONE THIRD CUP: Cut to size a **20 dram plastic vial** to the bottom of the closing nubs.

Masking tape label: *third cup.*

ONE HALF CUP: Cut a **40 or 60 dram plastic vial** to size. Use the template on page 83.

Masking tape label: *half cup.*

ONE CUP: Use a **60 dram standard cup**.

Masking tape label: *cup.*

PINT: Cut a **half-liter bottle** to size so it holds 2 standard cups, fair and full.

Masking tape label: *pint.*

TUB: Use a **one-pound soft margarine tub**. These often hold 3 standard cups, as manufactured. Check, and cut to size if necessary.

Masking tape label: *tub.* (If tubs of other sizes are also circulating in your room, designate the official status of this particular tub by underlining the label.)

QUART: Cut a **liter bottle** to size so it holds 4 standard cups when filled fair and full.

Masking tape label: *quart.*

grid paper (measure 4, 5, 6)

Photocopy the centimeter, inch, and lentil-sized grids on pages 52-54 in the recommended quantities. Trim around the outside dotted lines.

calculator (measure 5, 6)

Optional.

blunt scissors (measure 6, 8)

Sign for Basic Materials

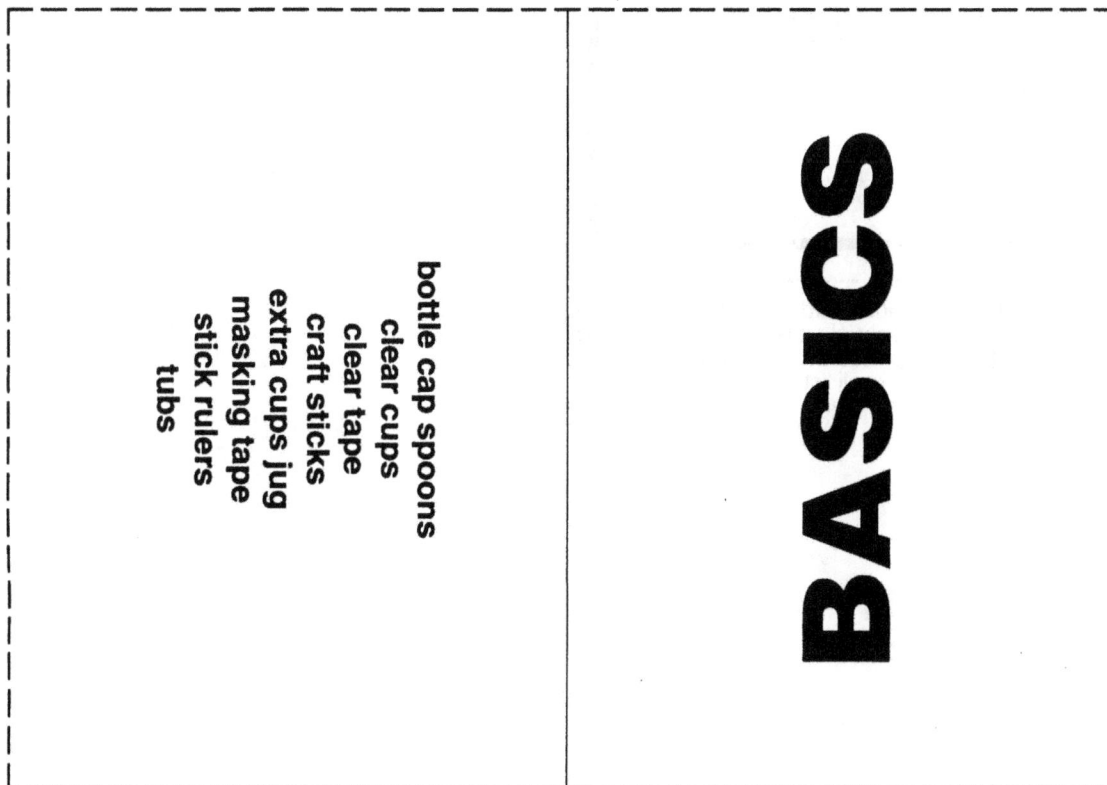

BASICS

bottle cap spoons
clear cups
clear tape
craft sticks
extra cups jug
masking tape
stick rulers
tubs

Cut and glue to a 4 x 6 inch index card. Fold it in half to create a stand up sign. Gather the listed materials in small quantities for storage next to this sign. Alternatively, cut this sign in half. Glue both pieces to a grocery bag that has been cut to size, or to a box of appropriate size, and store these materials inside. These *low - quantity basics* support Job Cards in two or more chapters.

MEASURE

6 containers

quart
tub
pint
cup
half cup
third cup

MEASURE

6 containers

quart
tub
pint
cup
half cup
third cup

Apply each label to a gallon storage jug with clear packaging tape. (Duplicate set recommended for large class sizes.)

grid paper
(measure 4,5,6)

set of 6 measuring cups
(measure 2, 3, 8)

C / MEASURE

special materials

CHAPTER SIGN: Glue to a 4 x 6 inch index card, and fold in half. Stand this sign in the space where you store special materials for this chapter. Or cut this sign in half, and glue both pieces to a grocery bag that has been cut to size, or to a box. Store all listed items inside.

Preparation: 1 copy

TEACHING NOTES

Purpose

To fill the same volume with different amounts of lentils. To recognize that lentils compress.

Introduction

▶ Fill a liter bottle to the top in the usual manner with scooper and funnel. Don't shake the contents.

▶ "Tickle" the sides of the bottle to settle the contents as much as possible. Discuss why the level of the lentils drops.

• Did any lentils leave the bottle? *No.*

• Why did their level fall? *They settled closer together and took up less space.*

▶ Right now all of us are packed rather loosely inside this classroom. How might we take up a lot less space? *Huddle together in a corner.* (Lots of fun.)

▶ Quickly fill 5 cups fair and full.

• Verbalize this 2-step process every time you fill a cup: overfill and shake once.

• Why shake only once? Fair and full is based on loose fill. You want to shake extra lentils off the top, but avoid settling the contents inside.

• Be sure to fill all 5 cups first, before pouring. Otherwise you may lose track of how many total cups you added to the bottle.

• Ask if 5 cups will fill the liter. Could the answer be both yes and no? Speculate. Build curiosity. But don't pour in the lentils.

Focus

◆ When you fill the bottle with a loose pack of lentils, will you shake it? *No. That would settle the contents.*

◆ When you fill the bottle with a tight pack of lentils, what will you do? *Shake it; tickle the sides; poke extra lentils into the top.*

Checkpoint

◆ How many cups of lentils filled the bottle? (How students answer this says something about their knowledge of and/or comfort level with fractions. Accept both whole and fractional definitions, pending their study of E/Divide.)

• Loose-pack: *less than 5 cups; 4 $9/10$ cups.*

• Tight-pack: *more than 5 cups; 5 $1/3$ cups.*

◆ May I see your write-up? (Look for these key terms: loose pack, tight pack, expanded, compacted, settled.)

◆ How did the lentils feel as you poked and pushed them into the top of the bottle? *(They felt*

---

c/1 You need...

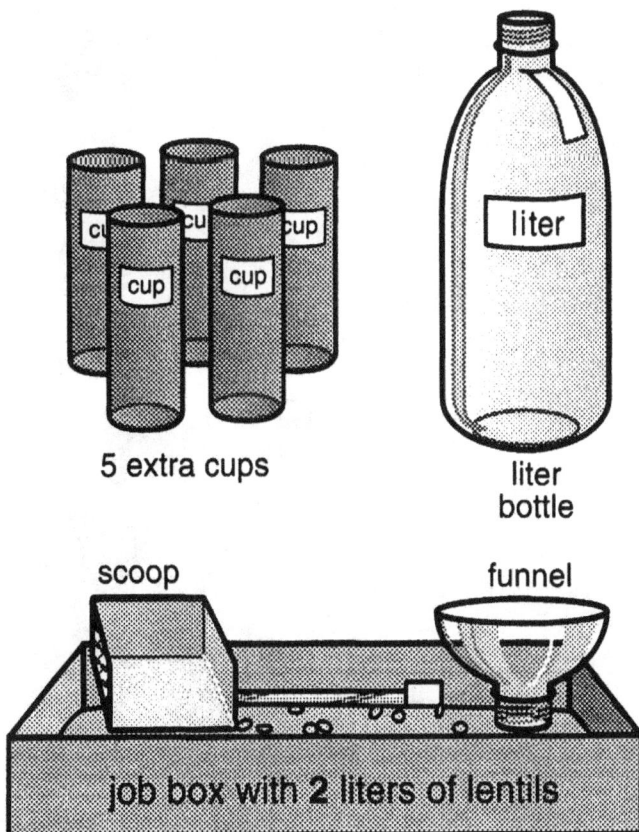

5 extra cups

liter bottle

scoop

funnel

job box with **2 liters of lentils**

measure 1   👤

How many cups fill the bottle?

LOOSE PACK?

TIGHT PACK?

FAIR AND FULL

Report your results. Why the difference?

TEACHING NOTES

Purpose

To pour lentils between 6 different measuring containers. To express equivalent volumes as equations.

Introduction

◗ Introduce the measuring set of 6 containers. Read the name on each one as you pull it from the half-gallon storage jug. Stand them in a row from smallest to largest: third cup, half cup, cup, pint, tub, and quart. (To create mathematical possibilities, we invented a new cup standard called the "tub." One tub equals 3 cups.)

◗ "Prove" that 3 half cups equal 1 cup. Create this discrepant event by filling the half cup only 2/3 full each time you pour it into the whole. Lead your class to correct this "error" by admonishing you to fill it fair and full every time.

◗ Confirm that 2 half cups really do equal 1 cup. Summarize this relationship as an equation. (Avoid abbreviating the name on the label. Write "half cup," not "1/2 cup.")

2 half cups = 1 whole cup

Focus

◆ Let me see you fill _____ fair and full.

◆ How many of these make one of those? Can you write this as an equation?

Checkpoint

◆ Twelve equations:

3 third cups = 1 cup	6 half cups = 1 tub
6 third cups = 1 pint	8 half cups = 1 quart
9 third cups = 1 tub	2 cups = 1 pint
12 third cups = 1 quart	3 cups = 1 tub
2 half cups = 1 cup	4 cups = 1 quart
4 half cups = 1 pint	2 pints = 1 quart

More

◆ Can you discover and write equations with a plus sign, so 2 *different* containers fill a third?

◆ Four quarts fill a gallon. Knowing this, can you extend your list of 12 equations to include gallons? (This is an exercise in thinking, not pouring lentils.)

C/2 You need...

jug of containers

job box with 2 liters of lentils

measure 2

How many smaller containers fill each larger container?

Can you discover and record 12 different equations?

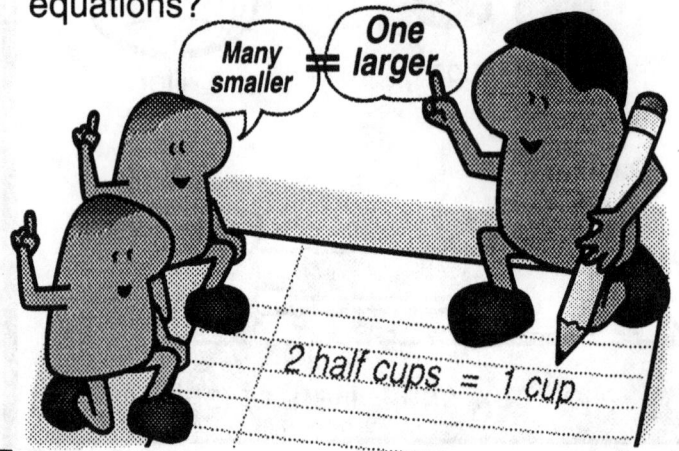

Many smaller = One larger

2 half cups = 1 cup

TEACHING NOTES

Purpose

To mark and label a liter bottle with six kinds of measure: quarts, tubs, pints, cups, half cups and third cups.

Introduction

▶ Demonstrate the quart calibration:

• Apply masking tape from the neck of the liter bottle to its base. Mark the baseline.

• Overfill the quart measure loosely and shake once in the usual manner. Then settle this fair and full measure vigorously, to facilitate funneling the lentils into the liter bottle without spilling.

• Invert the bottle for loose fill, and tip to level the contents. Mark the level with a line on the tape. Label it "qt" for quart.

▶ How shall we test the bottle to see if it really measures 1 quart? *Empty and refill it to the quart mark with a loose pack of lentils. Then see if this amount fills the quart fair and full.*

Focus

◆ How can you be sure the lentils are loosely packed in the bottle? *Invert the bottle. Tip the contents back to level.*

◆ How will you test the accuracy of your calibrations? *Refill to each mark and pour into each measuring cup.*

Checkpoint

◆ I see you have finished marking this bottle. Can you use it to measure ____? Do you think it will fill that container fair and full? Let's try it.

◆ Is it more accurate to measure by the bottle or by the cup? *By the cup.* (The end point of a cup is definite and easy to reach: simply fill it fair and full. The end point in a measuring bottle is less certain. A few lentils more or less register little height change in the wide bottle, yet produce a significant height difference when poured into the narrower cups.)

◆ Calibrated bottle.

More

Fill the measuring bottle with any unmeasured portion of lentils. Predict which container or combination of containers it will fill, and by how much.

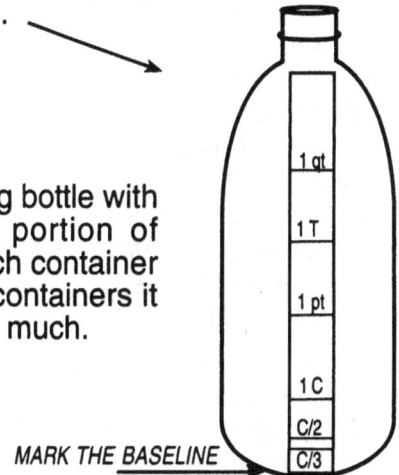

MARK THE BASELINE

C/3 You need...

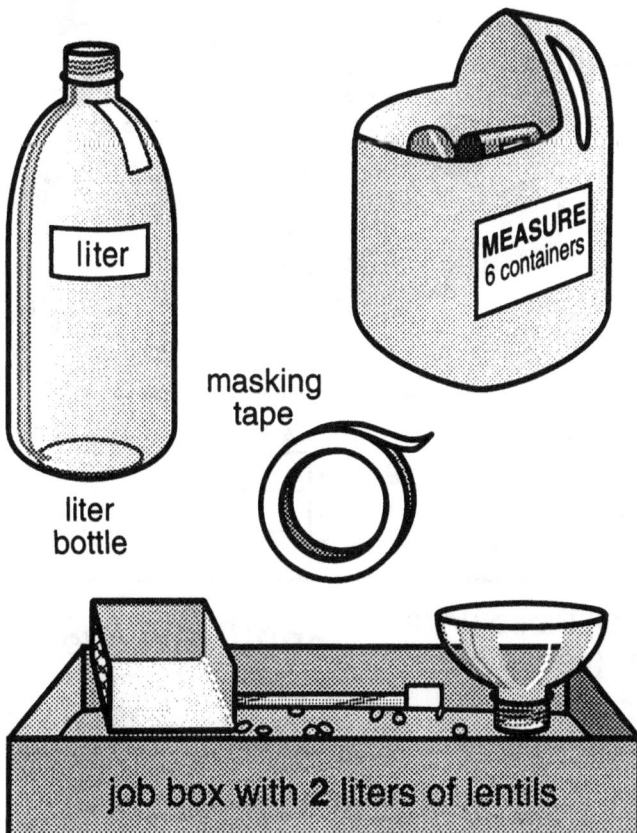

liter

masking tape

MEASURE 6 containers

liter bottle

job box with 2 liters of lentils

---

measure 3 👤

Stick masking tape on the liter bottle.

Mark and label how high the lentils reach.

LOOSE FILL

Quart reaches here.

BASELINE

abbreviations:
quart = **qt**
tub = **T**
pint = **pt**
cup = **C**
half cup = **C/2**
third cup = **C/3**

Pour both ways:

Cups ⤻ **Bottle**

Which measure is more accurate?

Draw the bottle full size. Stick the marked tape on it.

TEACHING NOTES

Purpose

To measure the length, width and height of a job box in centimeters, inches, and lentils.

Introduction

◗ Introduce centimeter units (*cm*):

• Display the cm ruler and grid.

• Think of objects about 1 cm long. *As wide as my fingernail. As thick as thin-sliced bread.*

• Measure the height of a third cup with the centimeter ruler. *Almost 6 cm.*

◗ Introduce lentil units (*len*):

• Scatter some lentils on the lentil grid. Notice how perfectly they fit in the squares.

• Examine the lentil ruler. How far do 20 lentils reach when lined up on the grid? *The length of the ruler; 20 lentil diameters; 20 len.*

• Is the diameter of a lentil 1 centimeter long? *No. A lentil is shorter than 1 cm.*

• Measure the height of a third cup with the lentil ruler. *A little over 9 len.*

◗ Introduce inch units (*in*):

• Display the inch ruler and grid.

• Think of objects about 1 inch long. *As wide as two finger nails. As thick as thick sliced bread.*

• Notice that the inch ruler and lentil ruler have equal length. So, 5 inches equal 20 lens. So, 4 lentils placed edge to edge span 1 inch.

• Measure the height of a third cup with the inch ruler. *A little more than 2¼ inches.*

Focus

◆ Which measure will you try first? *cm, len,* or *in.* (Students might measure in centimeter units today, then return to do lentils or inches tomorrow.)

◆ What will you measure with? *Grids and rulers.* (This makes the tables self-checking. Both ways of measuring should be in reasonable agreement.)

Checkpoint

◆ May I see your measuring tables?

◆ What units are you using? (Numbers without units mean very little. Are you measuring in paper clips, or worms, or what?)

◆ How accurately are you measuring? (Press more advanced students to use fractions or decimals. Inches should be written to the nearest 1/4 inch. Question data that is exactly the same in both columns. Every measurement carries experimental uncertainty. Students should express it honestly.)

C/4 You need...

2 grids each:

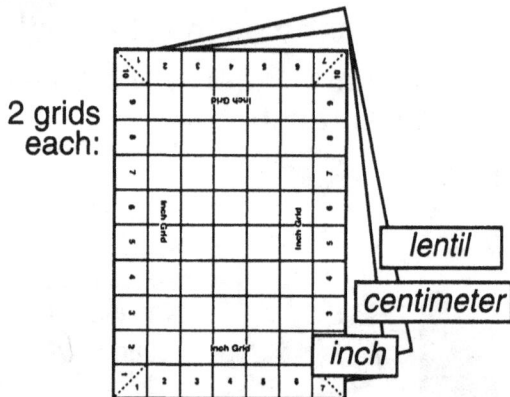

lentil
centimeter
inch

3 stick rulers

EMPTY job box

measure 4 🧍 or 🧍🧍

Measure these **lengths**...

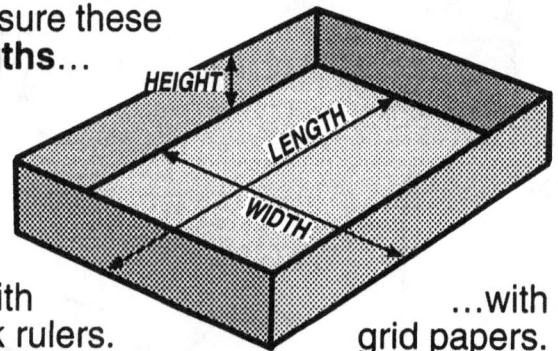

...with stick rulers.

...with grid papers.

Re-measure until you get good agreement..		RULER ANSWER	GRID ANSWER
	LENGTH		
	WIDTH		
	HEIGHT		

Try all these units:
  cm (centimeters)
  len (lentils)
  in (inches)

Length is a line!

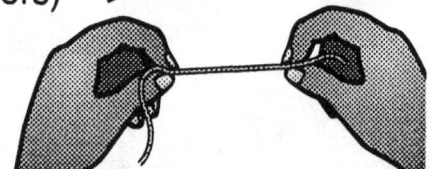

STICK RULERS

1. Copy and cut out on the dotted lines at least one ruler of each kind. (Cut the ends of each ruler precisely.)

2. Find a straight craft stick for each ruler you will make. Label the back of each stick with a pen: *centimeter ruler, inch ruler, lentil ruler.*

3. Lay a paper ruler and corresponding craft stick side by side on your table, with the ruler *face up* and the label on the craft stick *face down.*

4. Cut about 3 inches of clear packaging tape. Press one end of this strip to the center of the paper ruler. Pick up the ruler (now stuck to the tape) and center it on the blank side of the craft stick. The ruler will extend a little beyond each end of the stick.

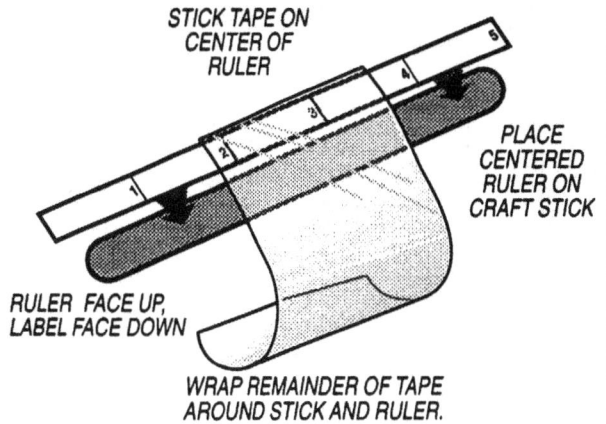

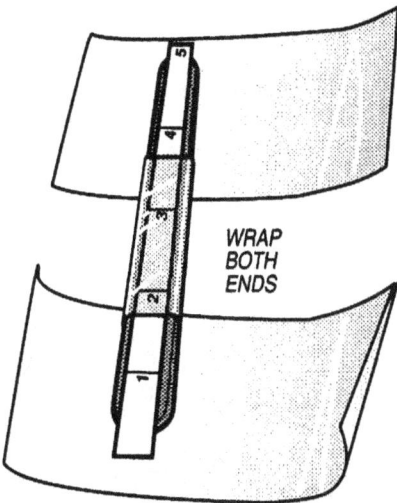

STICK TAPE ON CENTER OF RULER

PLACE CENTERED RULER ON CRAFT STICK

RULER FACE UP, LABEL FACE DOWN

WRAP REMAINDER OF TAPE AROUND STICK AND RULER.

5. Wrap the rest of the tape around the stick.

6. Wrap more tape around each end, completely sealing the paper. Use enough tape (up to 6 inches) to make each extension rigid.

7. Trim the tape carefully to each end of the ruler. You now have a durable Stick Ruler.

WRAP BOTH ENDS

One of each ruler needed in this and other chapters. You may wish to make extras if they will be used by many students.

centimeter rulers:

1	2	3	4	5	6	7	8	9	10	11	12
1	2	3	4	5	6	7	8	9	10	11	12
1	2	3	4	5	6	7	8	9	10	11	12
1	2	3	4	5	6	7	8	9	10	11	12

inch rulers:

1	2	3	4	5
1	2	3	4	5
1	2	3	4	5
1	2	3	4	5

lentil rulers:

1	2	3	4	5	6	7	8	9	10	11	12	13	14	15	16	17	18	19	20
1	2	3	4	5	6	7	8	9	10	11	12	13	14	15	16	17	18	19	20
1	2	3	4	5	6	7	8	9	10	11	12	13	14	15	16	17	18	19	20
1	2	3	4	5	6	7	8	9	10	11	12	13	14	15	16	17	18	19	20

centimeter grid

centimeter grid

centimeter grid

centimeter grid

Preparation: 5-20 copies

52

inch grid

inch grid

inch grid

inch grid

lentil grid

lentil grid

lentil grid

lentil grid

Preparation: 5-20 copies

Copyright © 1999 by TOPS Learning Systems

54

TEACHING NOTES

Purpose

To measure the area of the inside surfaces of a job box, in square centimeters, square lentils, and square inches.

Introduction

◗ Pantomime the difference between 1-, 2-, and 3-dimensional measure. Draw your class into this kinesthetic exercise so they can feel these different measures:

- **Length:** Pretend you are pulling invisible thread through your fingers.
- **Area:** Pretend you are polishing a large wall mirror.
- **Volume:** Pretend you are carrying a large armful of soccer balls.

◗ Demonstrate 2 ways to measure area:

- **Grid method:** Choose grids of the same unit. Completely cover any inside surface of the box and count squares. "Wallpaper" the sides with two grids folded in half and erected tent-style above the edge. Or "tile" the floor with 4 grids. (Students will usually count and multiply rows and columns in smaller areas, then add these subtotals to equal the whole. Neatness and organization are very important to minimize error.)

- **Ruler method:** Measure each side of the whole surface and multiply the numbers together. (Some students may remember to use their length measurements from the previous Job Card. This approach may require a calculator to multiply the larger numbers, and a review of decimals. Alternately, numbers can be rounded off, but areas will be very approximate.)

Focus

◆ Which measure will you try first? *sq cm, sq len, or sq in.*

◆ What will you measure with? *Grids and rulers.* (Once again, these tables are self-checking. With rounded-off measurements, agreement will be ballpark at best.)

Checkpoint

◆ May I see your measuring tables?

◆ What units are you using? (Numbers without units mean very little. Are you measuring in postage stamps, tortillas, or what?)

Special Note

◆ An actual lentil covers slightly less area than 1 sq len.

1 sq len
(3 times actual size)
lentil

C/5 You need...

4 grids each:

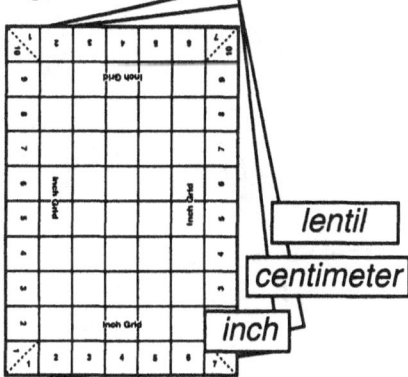

lentil
centimeter
inch

calculators are optional.

3 stick rulers

EMPTY job box

---

measure 5   🧍 or 🧍🧍

Measure these **areas**...

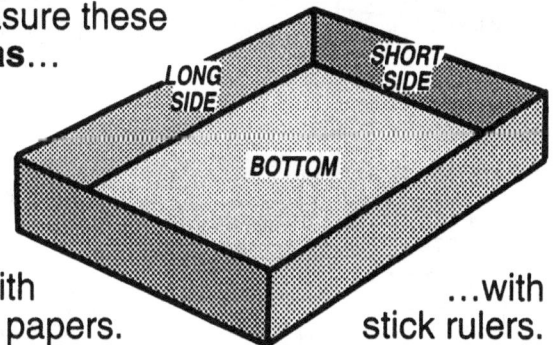

LONG SIDE   SHORT SIDE

BOTTOM

...with grid papers.

...with stick rulers.

Re-measure until you get good agreement.		GRID ANSWER	RULER ANSWER
	LENGTH		
	WIDTH		
	HEIGHT		

Try these units:
sq cm (square cm)
sq len (square len)
sq in (square in)

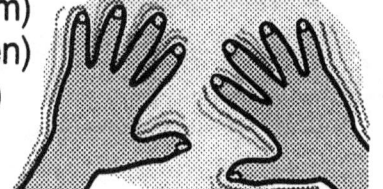

Area is flat!

TEACHING NOTES

Purpose

To measure the volume of boxes in cubic centimeters, cubic lentils and cubic inches.

Introduction

◗ Pantomime once again the difference between 1, 2, and 3 dimensional measure, as you did in the previous job card.

◗ Demonstrate how to cut, fold and tape boxes from grid rectangles. Use the Inch Grid for visibility.

• 1x4 inch rectangle: Fold this strip 4 times into a cube. Notice that it has no top and no bottom. This is called a *unit volume*, the unit being one inch.

• 4x10 inch rectangle: Fold this strip (across its 4-square width) on the 2nd, 5th and 7th lines to make a 4 x 3 x 2 inch box.

◗ Figure the volume of this larger box in 2 ways:

• Count the number of times the smaller cubic inch "room" fits inside the larger "hotel." Rest the hotel on an uncut inch grid, if helpful, to define the "floor plan." *Six rooms fit on ground floor of this 4 story hotel, making 24 room in all.*

• Multiply length by width by height. *2 in x 3 in x 4 in = 24 cu in.*

Focus

◆ How will you make each unit cube in the job card? *Cut a 1x4 area from each grid, fold and tape them into open-ended cubes.*

◆ Can you construct larger boxes? First make any sizes you like, then see if you can make the specific volumes detailed on the job card.

Checkpoint

◆ May I see the boxes you have made? How do you know that this box has this volume? (When finished, clip the boxes together, or tape them lightly to written work. The next job card requires these same boxes.)

◆ I see that many of your boxes are long and skinny. Can you make others that look more like bricks and cubes? (This question prompts children to expand volumes in all 3 dimensions at once.)

◆ Our Job Box measured:

49 cm x 37 cm x 8 cm = 14,500 cu cm
76 len x 58 len x 12 len = 53,000 cu len
19 in x 14.5 in x 3 in = 830 cu in

C/6 You need...

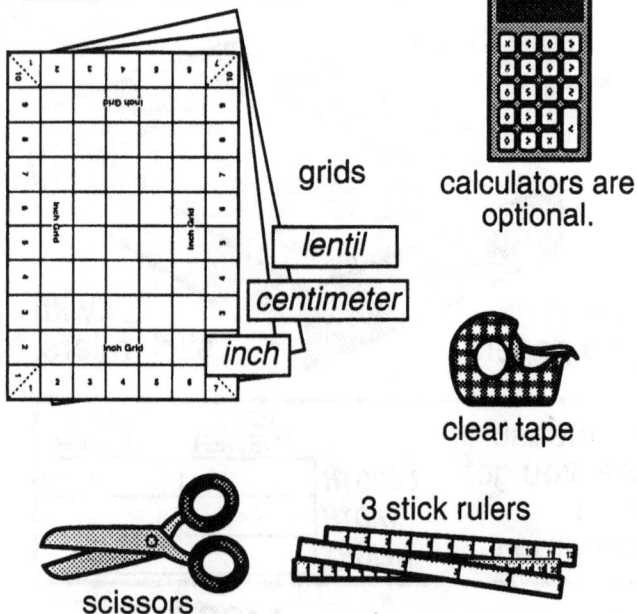

grids

calculators are optional.

lentil

centimeter

inch

clear tape

3 stick rulers

scissors

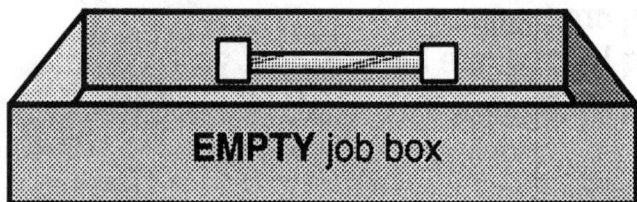

EMPTY job box

---

measure 6   🧍 or 🧍🧍

Volume takes up space!

Cut, tape and label these **unit volumes**:

4 sides, no top, no bottom.

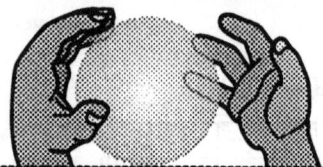

1 cu in

1 cu cm

1 cu len

Also make bigger boxes from each kind of grid. Label the volumes on each box.

Cut, tape and label these volumes:

2 cu in	*3 cu cm*	*4 cu len*
4 cu in	*9 cu cm*	*16 cu len*
8 cu in	*27 cu cm*	*64 cu len*

Measure the volume of your job box in all these units: *cu cm, cu len, cu in.*

TEACHING NOTES

Purpose

To graph the distribution of lentils dropped on a target. To see a relationship between variables.

Introduction

▶ How to hit the center of the target:

• Hold the Stick Ruler vertically beside the bull's-eye. Hold the Bottle-Cap Spoon fair and full, at a chosen height on the stick. The lip you pour from should be positioned directly over the bull's-eye, so the lentils hit dead center as you tip the cap.

• Pour slowly. Fast flips fling lentils too far out.

▶ How to count:

• Start at the outside ring and work in.

• Sweep lentils off the paper as you count them.

• Move "liners" into the nearest ring.

Focus

◆ How many lentils do you need to fill the Bottle-Cap Spoon? *Less than a handful.* (A small pile in a corner of the Job Box will suffice.)

◆ What units of measure will you use? (Any measure is OK. If students wish to compare results, they may need to convert, which is a good lesson in math. Or they may need to agree on a standard, which is a good lesson in politics.)

Checkpoint

◆ May I see your graphs? If I divided this page into 4 little graphs, left out the drop-heights, and scrambled them, could you order them by drop-height? What logic would you use?

◆ May I see your written work? (Low drops concentrate lentils in the 3 or 4 center rings. Medium drops create bell-shaped distribution patterns, centered about the middle rings. High drops create low, flat patterns, with most lentils bouncing away from the center and outside the rings.)

More

◆ Scatter Puzzles:

1. Choose a drop height. Write it on scratch paper and hide this number.

2. Drop a bottle cap of lentils from this height onto the center of the target. Mark each lentil position with an "x."

3. Trade scatter puzzles with a friend: Can you say what the drop height is not? Might be? Probably is? Offer a range of values.

◆ Read about meteors and impact craters. How is the size of the rock related to the size of the hole? (Materials: a tarp, a box of lentils or a sandbox, a meter stick, a set of measuring cups, a variety of rock sizes, flat dry ground.)

C/7 You need...

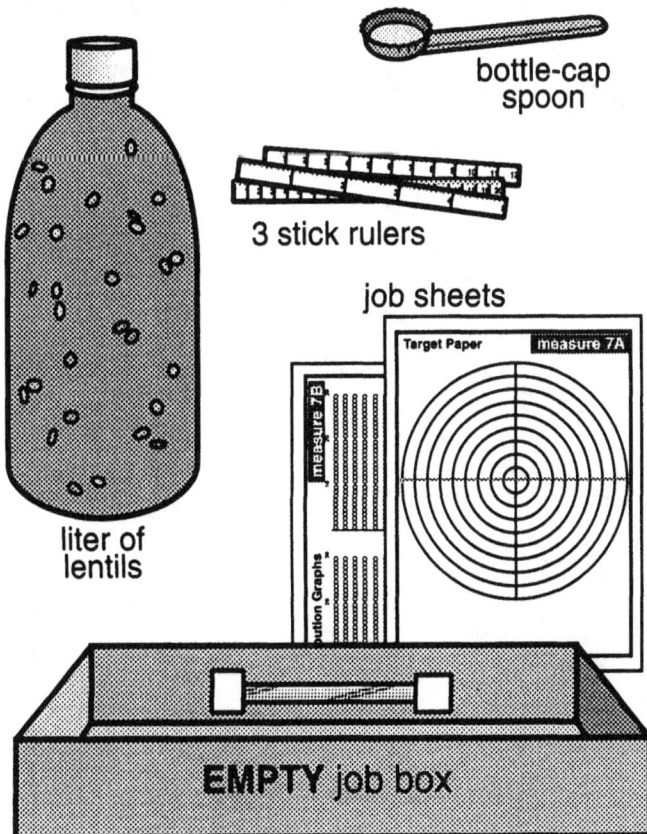

bottle-cap spoon

3 stick rulers

job sheets

Target Paper — measure 7A

measure 7B

Distribution Graphs

liter of lentils

EMPTY job box

---

measure 7 👤 or 👤👤

Drop a fair-and-full spoon of lentils from different drop heights onto the target.

Drop Height: 3 inches. Ready!

Starting my pour. Slow and steady...

Target

Graph the scatter pattern. What trends do you notice?

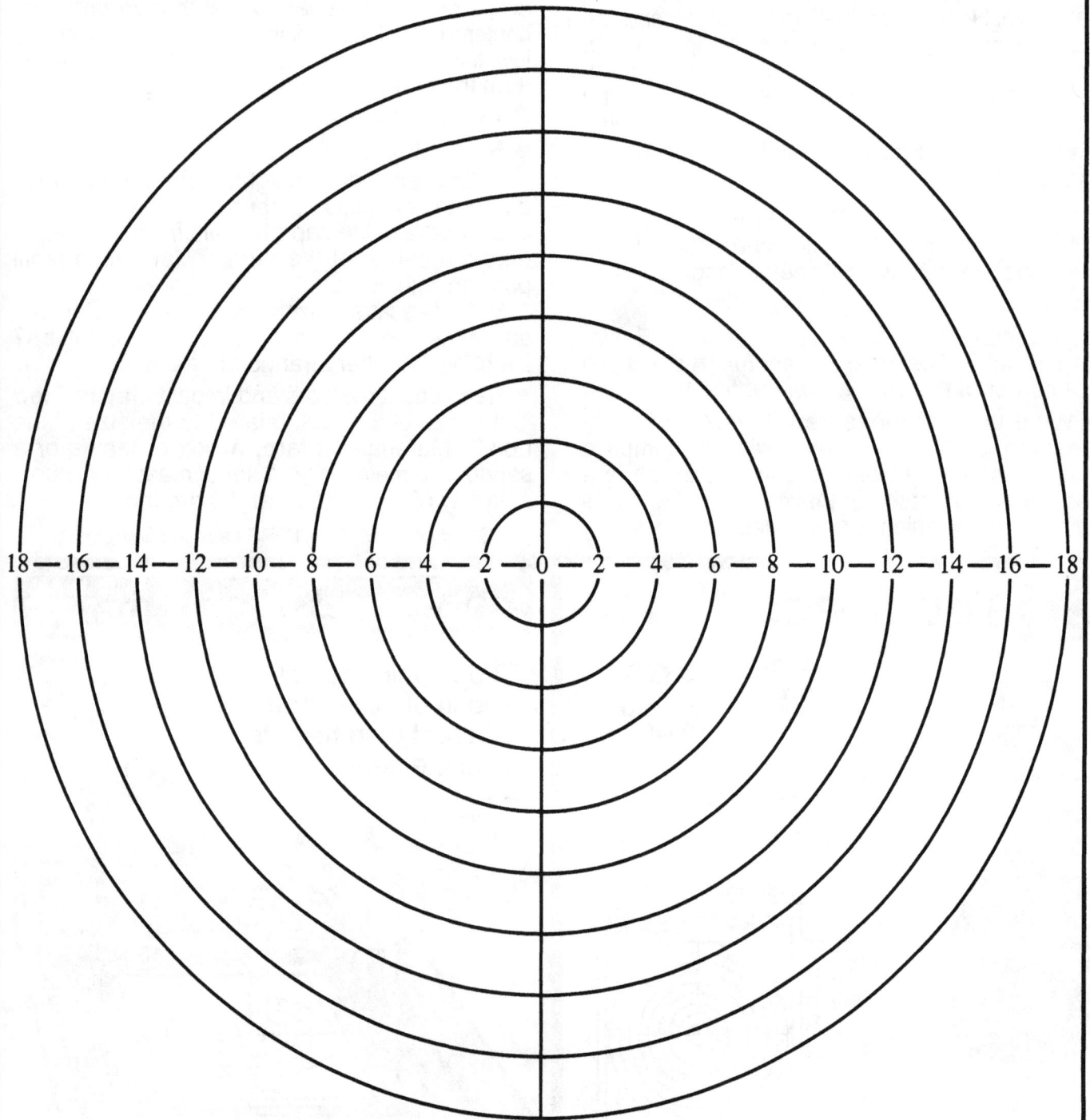

Target Paper

18 — 16 — 14 — 12 — 10 — 8 — 6 — 4 — 2 — 0 — 2 — 4 — 6 — 8 — 10 — 12 — 14 — 16 — 18

Job Sheet: several copies

Distribution Graphs

drop height = [ ]

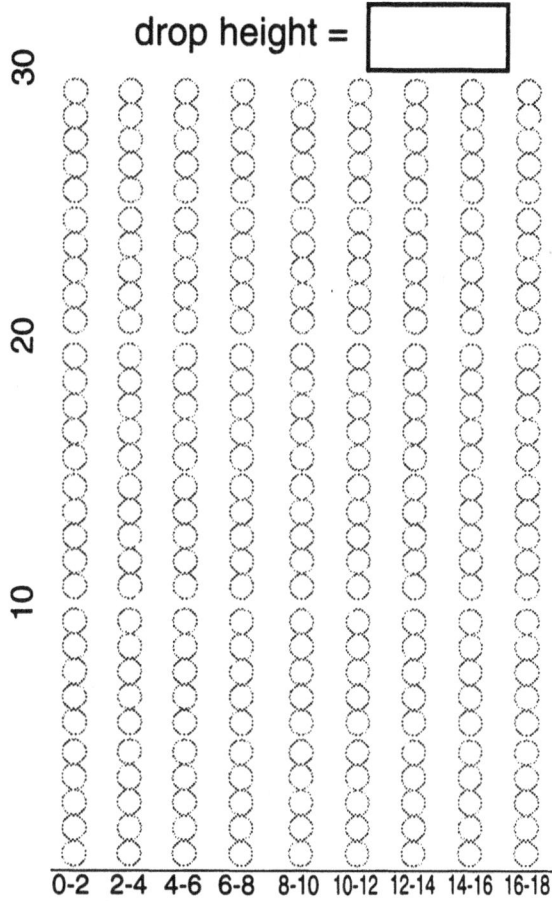

0-2 2-4 4-6 6-8 8-10 10-12 12-14 14-16 16-18

drop height = [ ]

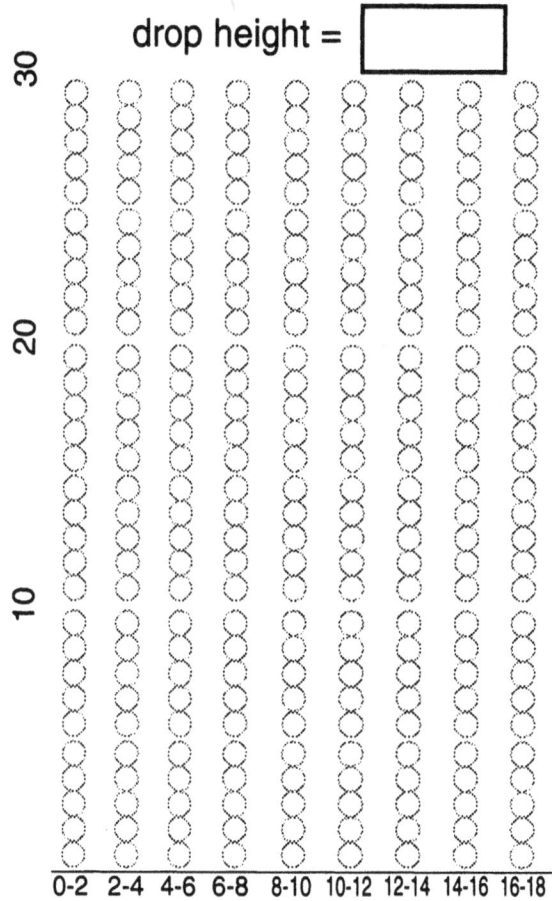

0-2 2-4 4-6 6-8 8-10 10-12 12-14 14-16 16-18

drop height = [ ]

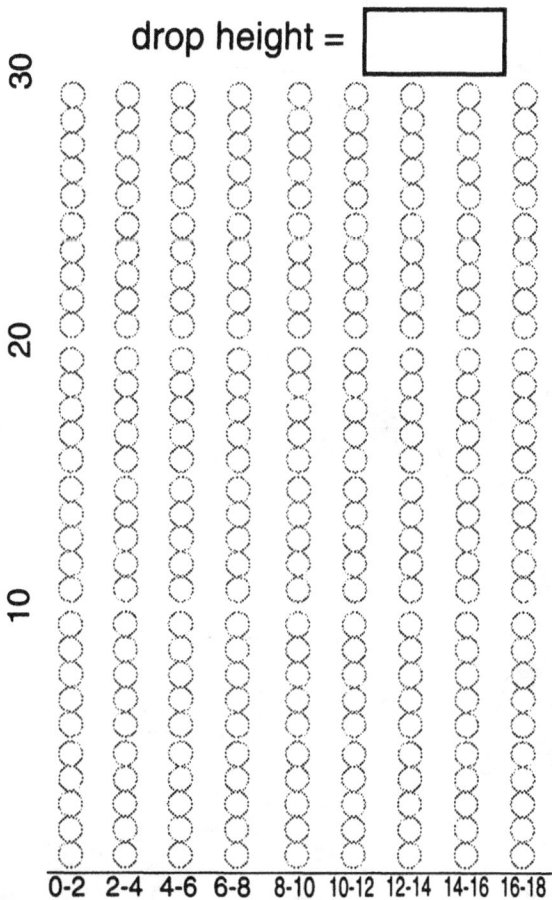

0-2 2-4 4-6 6-8 8-10 10-12 12-14 14-16 16-18

drop height = [ ]

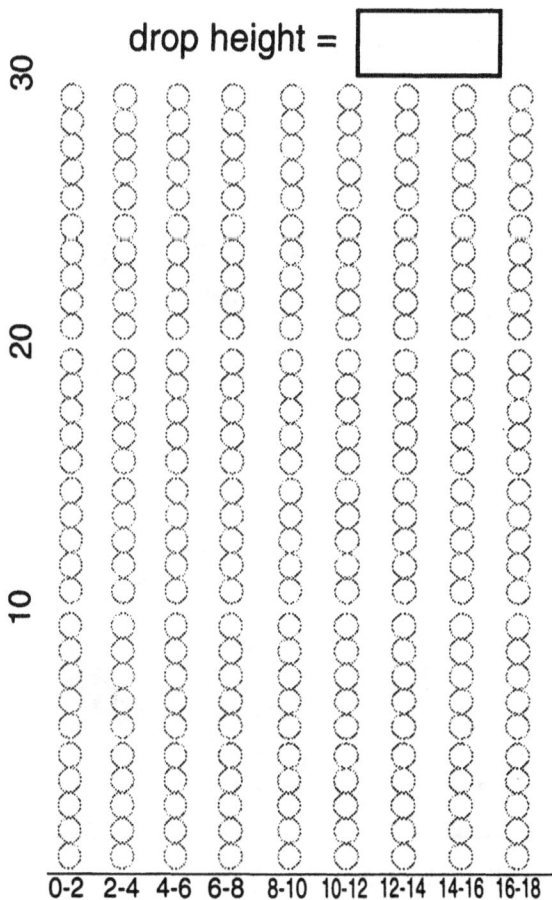

0-2 2-4 4-6 6-8 8-10 10-12 12-14 14-16 16-18

Job Sheet: 1 copy per student

LESSON NOTES

Purpose

To provide students with an approved way to pursue their own ideas about mmeasuring.

Introduction

Display this card when you want to do the suggested activity or try you own experiments.

Focus

◆ What do you want to study about measuring?

◆ Are these the materials you need?

Checkpoint

Report in words and pictures:
- What you did.
- What you learned.
- Questions you may still have.

Special Note

Any great ideas? Record them on the opposite page.

C/8 You may use...

MEASURE
6 containers

job box with **2 liters of lentils**

☞ Ask your teacher for other items.
Tell why you need them.

---

measure 8 👤 or 👥👥

On your own.

My idea:
I will make my own set of paper measuring cups, and explore their advantages and limitations.

These cones seem to work well.

cup

half

D / DESIGN

In this chapter: Create "wild" places out of rock, cardboard and lentils. Build a mountain, perhaps, or a mesa. Think of your hands as the wind and the rain. Erode the landscape. Sculpt rivers into lakes, deltas into oceans. Meanwhile, back in civilization, Lentilville continues to experience growing pains. City planners are talking about relocating City School north of Lentil River and building a new bridge at NE 7th Avenue and Dogwood for city access. Across town, happy inventors are busy creating flip chart movies and designing hourglass clocks to run at variable speeds.

Basic Materials: Quantities define maximums needed to support any one Job Card in this chapter. Store *high-quantity basics* (Job Boxes, liters of lentils, bottle lids, scoops, funnels) on and under a table or counter. Store *low-quantity basics* near the "basics" sign (see page 45) or in a "basics" box. Consult our Glossary on pages 6-9 for a full description of these items. See the next page for additional special materials used in this chapter.

- [ ] **1 job box**
- [ ] **up to 3 liter of lentils**
- [ ] **1 scoop**
- [ ] **1 funnel**
- [ ] **50 craft sticks**
- [ ] **masking tape**
- [ ] **1 tub**

☞ *Please observe our copyright restrictions on page 2.*

Store these chapter-specific items together in a designated place. They require about 1 square foot of dedicated space. General classroom materials (like scissors and tape) are also listed below when used, while others (like pencil and paper) are always assumed.

landforms book (design 1)

Photocopy and assemble the <u>booklet</u>.

set of 3 rocks and 3 cards (design 1, 2, 3, 9)

Flat rocks, ranging from hand size to palm size, with angles are best. Cut a piece of **corrugated cardboard** to roughly match the shape of each rock. Store in a <u>**gallon storage jug**</u>.

Paper label(s) provided.

landscape props design 2, 3)

1. Make one photocopy each of the landscape/townscape line masters. (Find these on pages 67, 70 and 71.) Set the street sticks labels aside.

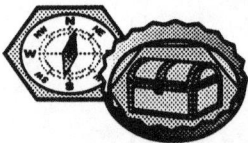

2. Cut each strip of props along the zigzagged edges. Cut out the 3 hexagons at the end of one landscape strip; press each one into a **bottle cap**.

4. Crease the remaining strips along the 3 long dotted lines before you separate them. Fold back on the center line so graphics show on both sides, then fold up the pointed grey tabs.

5. Cut the strips apart on the solid lines to make separate stand-up figures, and press their grey bases into bottle caps. No glue is needed.

6. Store the props in a **clear cup**. Label a **clothespin** "N" on both wings and add this to the cup.

Paper label provided.

townscape props (design 4, 5)

Store the townscape props in another **clear cup**.
Paper label provided

street sticks (design 4, 5)

These labels for 16 avenues and 15 streets were already copied (above). Cut out each double name (31 rectangles) and glue them to 31 **craft sticks**:

1. Rub or brush a film of **white glue**, or use a glue stick, on both sides of a craft stick, stopping just short of each rounded end.

2. Holding the stick by the ends, press a sticky edge along the center of a street label lying *face down* on your table.

4. Pick it up with the street label stuck to it, and turn it over. The grey line on a flat "roof" is now centered over a slightly longer "wall."

5. Rest the bottom edge of this wall on your table and press the roof around both sticky sides. Only each end of the stick and its bottom edge remain uncovered.

When the glue is dry, bundle the street sticks in a rubber band.

butcher paper (design 6)

Use a roll that is at least 3 feet wide. You may wish to precut 5 foot lengths.

meter stick or yard stick (design 6)

blunt scissors (design 7, 9)

double lid (design 8)

Slice the "roofs" off two **liter bottle lids**, leaving two threaded tubes. Do this by hand with a **hacksaw** (or electric band saw.) Tightly screw each lid onto a liter bottle to provide a safe grip. Cut into the top edge of each cap by only the thickness of the blade. The bottle's rim helps guide the saw blade, keeping you from shaving off too much.

Join cleanly cut caps, cut edges together, with a single layer of masking tape. Then overlay with a long strip of **electrical tape**, stretched and wound very tightly several times to make a solid connection. (Note: Without a masking-tape base, the tight electrical tape may force the caps to slip off-center over time.)

Such connectors also come ready-made. Look for "tornado tubes" sold in science supply catalogs. Or you might make a permanent hourglass by joining them directly with electrical tape over a masking tape base.

straight pin (design 8)

This is used with the double lid above. It should be no longer than 1 inch for safety reasons. (When you poke it between the plastic lids it must not stick out the other side.) Use a **cork** or a piece of labeled cardboard as a pin cushion to keep the pin from getting lost.

canning ring (design 9)

The smaller standard size works best.

landforms book (design 1)
set of 3 rocks and 3 cards (design 1, 2, 3, 9)
landscape props (design 2, 3)
townscape props (design 4, 5)
street sticks (design 4, 5)
butcher paper (design 6)
meter stick (design 6)
double lid (design 8)
straight pin (design 8)
canning ring (design 9)

DESIGN
3 rocks
3 cards

DESIGN
3 rocks
3 cards

Apply to a <u>gallon storage jug</u> with clear packaging tape. (Extra label provided in case you need a duplicate set to serve many students.)

D / DESIGN
special materials

CHAPTER SIGN: Glue to a 4 x 6 inch index card, and fold in half. Stand this sign in the space where you store special materials for this chapter. Or cut this sign in half, and glue both pieces to a grocery bag that has been cut to size, or to a box. Store all listed items inside.

LANDSCAPE
props
bottle cap figures
"N" marker

TOWNSCAPE
props
bottle cap figures

Apply each sign to a <u>clear cup</u> with clear packaging tape.

Preparation: 1 copy

TEACHING NOTES

Purpose

To model a variety of geographic features in a lentil box.

Introduction

▶ Consider some of the geographic diversity that graces our home planet. Leaf through the Landforms Book. Read the titles on each page, and the related terms underneath.

▶ Demonstrate how to model the delta on page 7 of the Landforms Book:

• It recommends 1 liter of lentils, so pour that amount into an empty Job Box. Spread the lentils across the bottom and into corners. A quick, sharp shake creates uniform cover.

• Create the landform in concept, but don't copy the exact form. Make more or less ocean; run the river in a different direction with more or fewer channels.

• Draw top and side views on your blackboard. Don't get lost in detail. Draw a dashed line across the top view to define the cutaway side view.

• Ponder the geohistory by asking questions. How did this delta form? Why is it so flat? Why does the river bend and braid? How might this delta continue to change? Where might I search for more information?

Checkpoint

◆ May I see the top and side views you drew? (Conceptual accuracy? Good labels?)

◆ May I see your written work? (Are attempts at geohistory – erosion, uplift, folding, faulting – part of the explanation?)

◆ Gently push your students toward higher, more complex levels of thought and expression:

• I see you made a watercourse. Is this a creek or a river? Where did this water come from and where is it going?

• I see you made an archipelago. How would this chain of islands appear to a fish? Are these islands volcanic?

• I see you have made a lake. Is it deep or shallow? How will it change as it ages?

More

◆ Additional landforms to model:

• An actual landform in your vicinity.

• A glacier of flowing lentil "ice" between rocks.

• A prairie (steppe, tundra, savanna, desert.)

◆ Make an illustrated Landforms Dictionary. Start with related terms in the Landforms Book. Pronounce and define each word. Sketch an example. Find as many additional terms as you can. (This could make an excellent class project.)

D/1 You need...

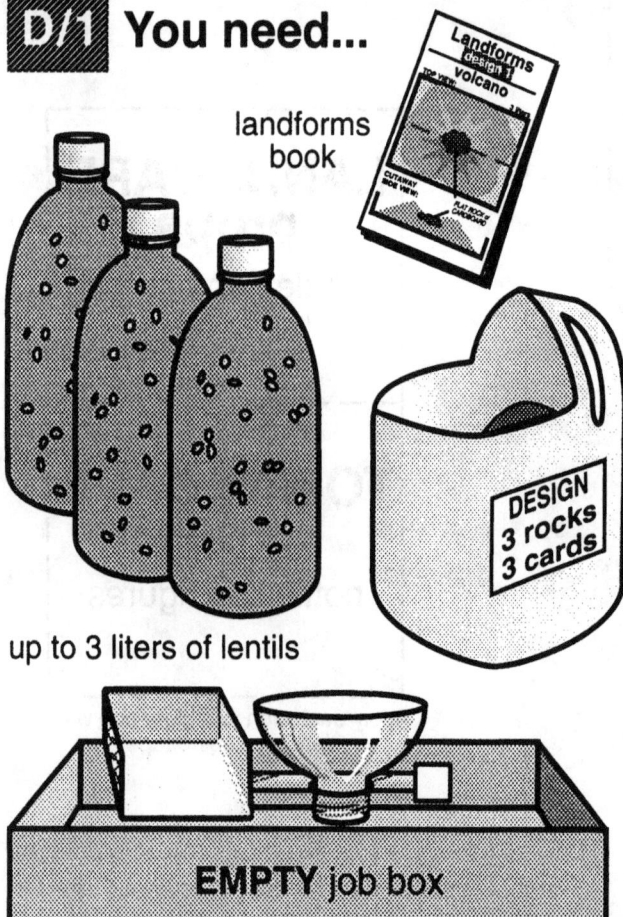

landforms book

up to 3 liters of lentils

EMPTY job box

DESIGN
3 rocks
3 cards

---

design 1   ♦ or ♦♦

Think of your hands as wind and rain.

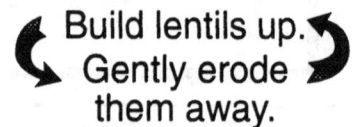

Build lentils up.
Gently erode them away.

Look to the Landforms Book for ideas.

Name, draw, and describe. How did natural forces create these forms?

Landforms
design 1

1. volcano — 3 liters

TOP VIEW:

FLAT ROCK or CARDBOARD

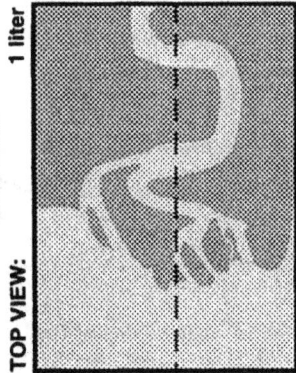

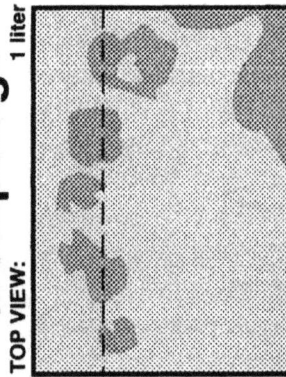

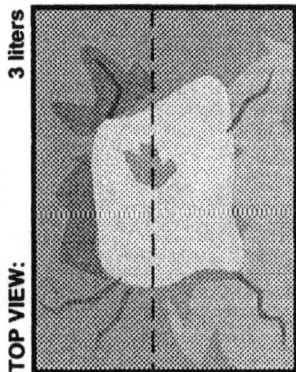

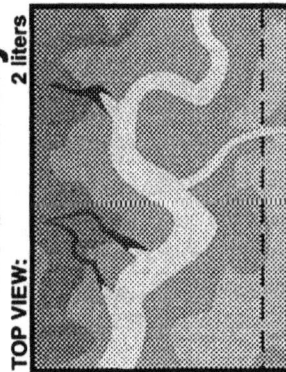

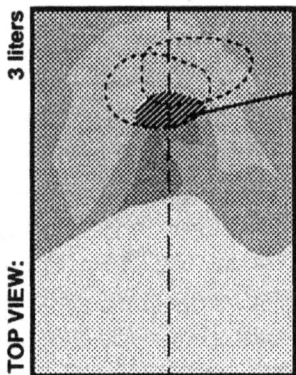

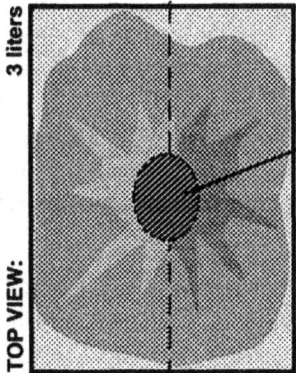

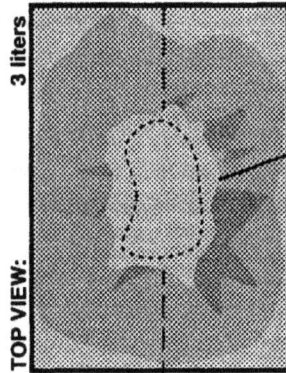

CUTAWAY SIDE VIEW:

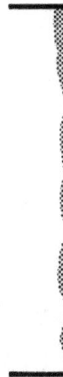

Related Terms: crater (lake), lava dome, cone, caldera, vent, peak, prow.

2. mesa — 3 liters

TOP VIEW:

FLAT ROCK or CARDBOARD

CUTAWAY SIDE VIEW:

Related Terms: plateau, butte, cliff, crag, ridge, rockslide, talus slope.

3. cliffs / caves — 3 liters

TOP VIEW:

FLAT ROCKS or CARDBOARD

CUTAWAY SIDE VIEW:

Related Terms: cavern, grotto, fault, glacier, erratic, stalactite, stalagmite.

4. mountains / lakes — 3 liters

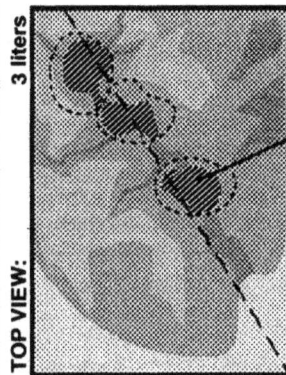

TOP VIEW:

FLAT ROCK or CARDBOARD

CUTAWAY SIDE VIEW:

Related features: stream, river, waterfall, spring, glacier, moraine.

5. inland sea — 3 liters

TOP VIEW:

CUTAWAY SIDE VIEW:

Related Terms: basin, bog, moor, pond, drainage, oasis.

6. river valley — 2 liters

TOP VIEW:

CUTAWAY SIDE VIEW:

Related Terms: canyon, gorge, flood plain, *tributary, arroyo,

7. delta — 1 liter

TOP VIEW:

CUTAWAY SIDE VIEW:

Related Terms: alluvial fan, bayou, estuary, swamp, wetlands, marsh.

8. archipelago — 1 liter

TOP VIEW:

CUTAWAY SIDE VIEW:

Related Terms: ocean, island, gulf, bay, cape, cove, reef, shoal, fjord, peninsula, strait, isthmus, continent.

Preparation: 1 copy

TEACHING NOTES

Purpose

To model, draw and write about a wonderful wilderness landscape. To exercise creativity and imagination.

Introduction

♦ Imagine that you are taking a vacation in a wild place, far away from people. Look through the Landforms Book for inspiration and ideas. Decide as a class where you would like to visit. Then "go there" by arranging lentils in a job box.

♦ Carve a stream or lake or beach into the lentil "land mass." Ask what other kinds of things you might see, or smell, or hear, or touch, or taste in this wild environment. Brainstorm a list of natural objects.

♦ Add bottle-cap Landscape Props to your lentil wilderness from this list. Draw an original prop, as well, on folded scratch paper. Stand it like a tent in your lentil wilderness.

♦ Improvise a class story about the special place you have created. Open with a sentence like, "Once upon a time our class went on a summer vacation...." Or begin with the classic, "It was a dark and stormy night...." Ask volunteers to each contribute an additional sentence that builds a story chain. Call forth the naturalist in your students by prompting them to describe in rich detail the worlds they see in their imaginations.

Focus

◆ What kind of a wild place will you create? Which landscape props will you use? Will you include props that you make yourself? (Original props may be taped or glued later into illustrated stories.)

◆ What kind of an adventure will you write about? Can you organize your thoughts in an outline?

Checkpoint

◆ Tell me about the wild place you created.

◆ May I hear/read about your adventures?

More

◆ Create designs in negative space: Shake one liter of lentils smooth in the bottom of a Job Box to begin. Scrape away the "frost" on the "window."

◆ Create designs in positive space: Begin with an empty Job Box and one full liter of lentil "paint."

D/2 You need...

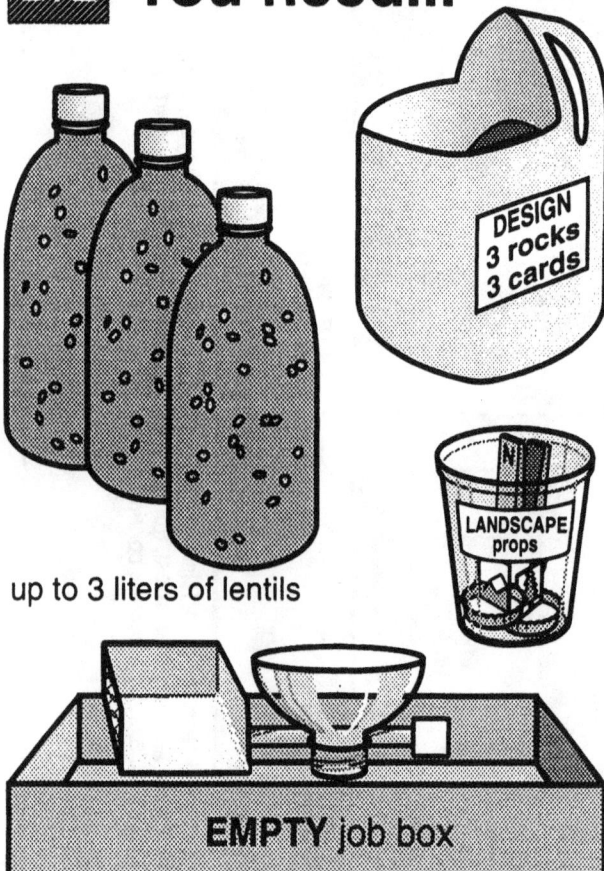

up to 3 liters of lentils

EMPTY job box

design 2   👤 or 👥👥

Create a wonderful wild place.

Draw and write about your adventures.

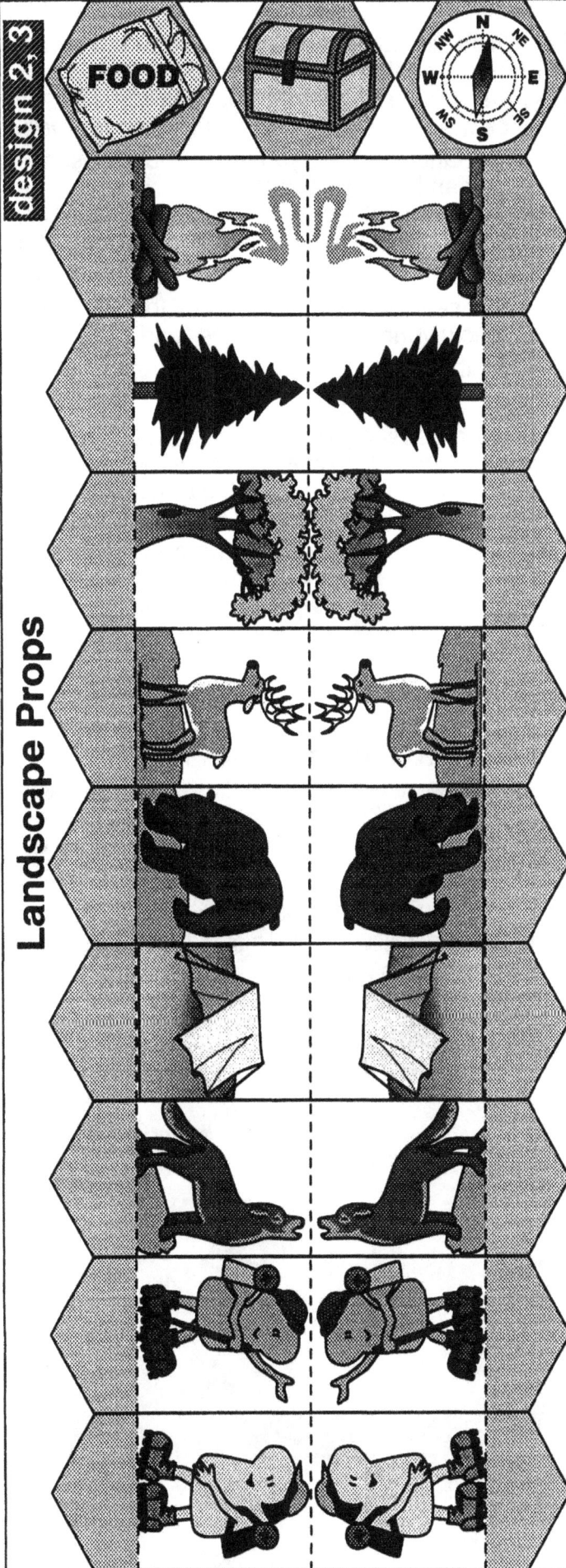

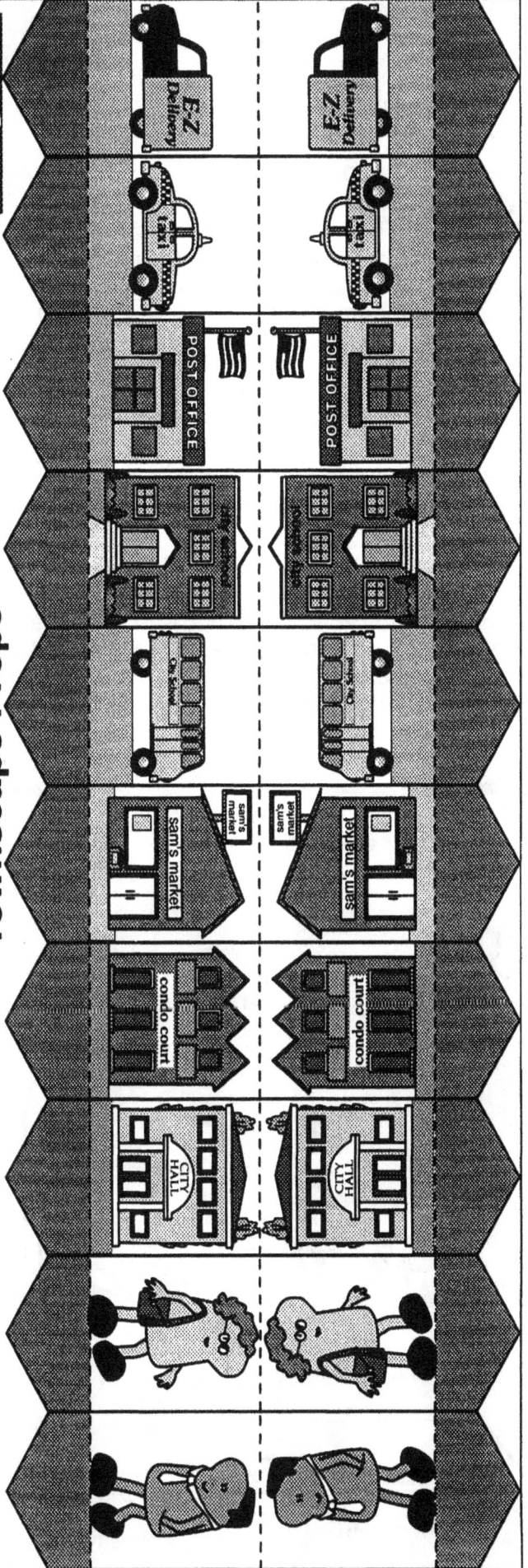

FOOD

Landscape Props

Townscape Props

E-Z Delivery

taxi

POST OFFICE

On School

sam's market

condo court

CITY HALL

67

Preparation: 1 copy

TEACHING NOTES

Purpose

To model and map a landscape with geographic reference points.

Introduction

◗ Create a simple landscape, with a lake or stream, in the job box. Define north by attaching the clothespin labeled N to any side of the box.

◗ Set the "compass" in this landscape so N, S, E and W line up with the sides of the box. Notice that N on the compass does not usually point to N on the clothespin. The clothespin represents geographic north, a point thousands of miles away. Mention, too, that your compass needle is pointing to magnetic north, which doesn't quite line up with geographic north. (This is true at most locations around the globe.)

◗ Bury the bottle-cap "food cache" anywhere in the box, near a "tree," perhaps, or other defining landmark. Place a bottle-cap "hiker" elsewhere in the landscape.

◗ Model on your blackboard how to write a set of directions that lead the hiker to the food cache (around a watercourse, perhaps). Include cardinal points in your description, and a simple drawing.

Focus

◆ Which side of your Job Box will you tag north? How will you line up the Bottle-Cap Compass? *So the cardinal points line up with the sides of the box, and N points to the side tagged north.*

◆ What landmarks will you put in your landscape so you can have good reference points for your map? (Remind students to show you their maps, if possible, before taking apart their landscapes.)

Checkpoint

◆ May I see your map? Let's see if this little hiker can follow your directions to the food cache.

◆ Was this map accurate? Why do you think so?

More

◆ Bury a treasure within a landscape. Draw an accurate top view, to scale, so someone else can use your map and easily find the treasure without tearing up the landscape.

◆ Research the terms magnetic pole and declination. Draw a sketch to explain what you learn. Find out the magnetic declination for your town.

D/3 You need...

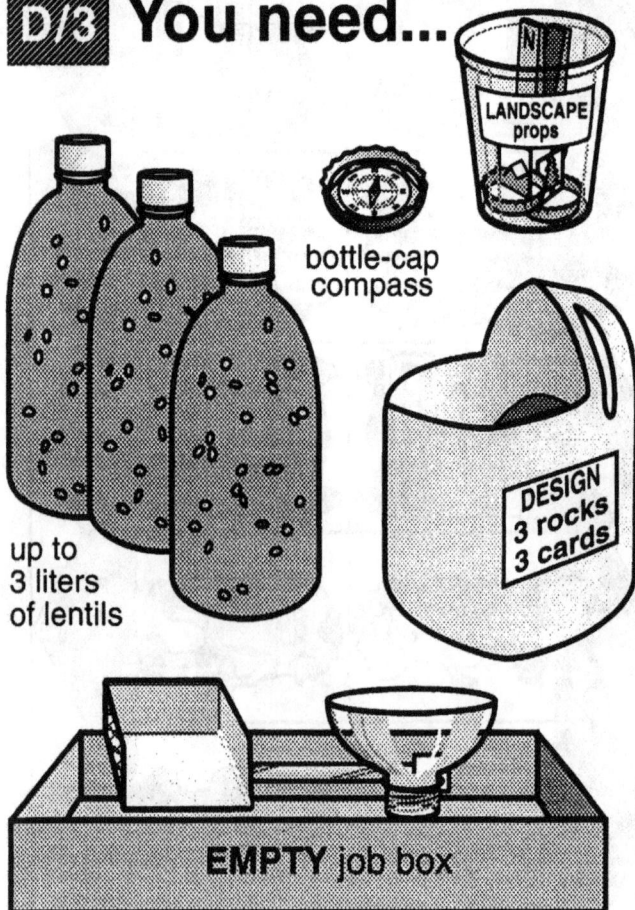

bottle-cap compass

LANDSCAPE props

DESIGN 3 rocks 3 cards

up to 3 liters of lentils

EMPTY job box

design 3 ♀ or ♀♀

Scenario: You are leading an extended pack trip into the wilderness. To travel more lightly, you decide to cache a supply of food to pick up on your return trip. Hang it out of reach of wild animals.

Draw a reliable map. Your survival depends on it.

Cache is on tall rock, 30 paces SE of lone pine, at southward bend of stream.

PINE

FOOD CACHE

TEACHING NOTES

Purpose

To lay out a city grid in logical order. To practice giving directions in terms of right and left, cardinal points, and street names.

Introduction

▸ Introduce yourself as the mayor of Lentilville. Welcome your class to town. As you present each section below, sketch the corresponding part of this map to give your students a basic orientation:

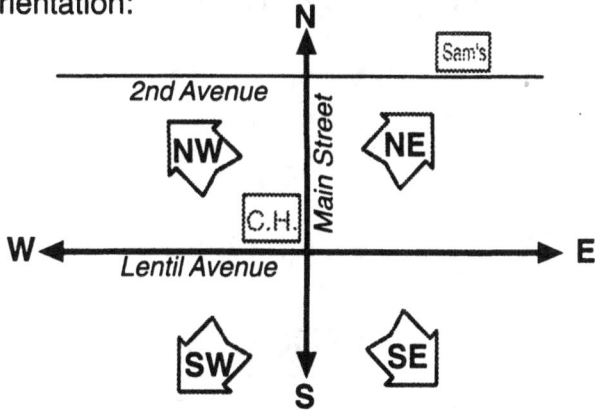

• City Hall (C.H.) is located in the center of town, at the corner of Main Street and Lentil Avenue.

• All streets (including Main Street) run N-S; all avenues (including Lentil Avenue) run E-W.

• Main and Lentil divide the town into four quadrants: NE, SE, SW, and NW.

• Sam's market is located on 2nd avenue NE, on the north side of the street, a little off Main Street. Who can draw Sam's market on my map?

• One of you walks into City Hall, and asks a clerk for directions to Sam's market. Who can play the part of the clerk? *Go out the front door and turn left, to the corner of Lentil and Main. Head north up Main. Turn right on 2nd, heading east. Sam's Market will be on your left.*

Focus

◆ Sort the Street Sticks into like piles. / Lay out Lentil and Main in the right order. / Add the rest of the streets and avenues.

◆ You must place City Hall at the center of town. Where will you put the rest of the props?

◆ Can you write directions for getting from ____ to ____?

Checkpoint

◆ Show me the directions you have written.

◆ Tell this "peoplet" how to get from ____ to ____.

◆ If I am heading ____(N/S/E/W), should I turn right or left to head ____(N/S/E/W).

D/4 You need...

street sticks

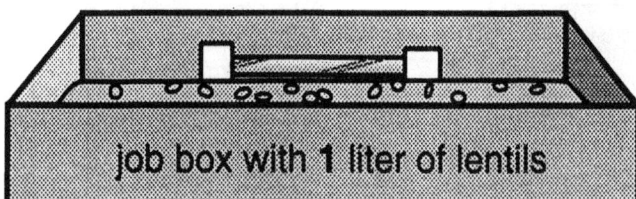

job box with 1 liter of lentils

---

design 4 👤 or 👥

Lay out the city of Lentilville in a logical sequence of avenues and streets. Begin like this:

Corner of job box.

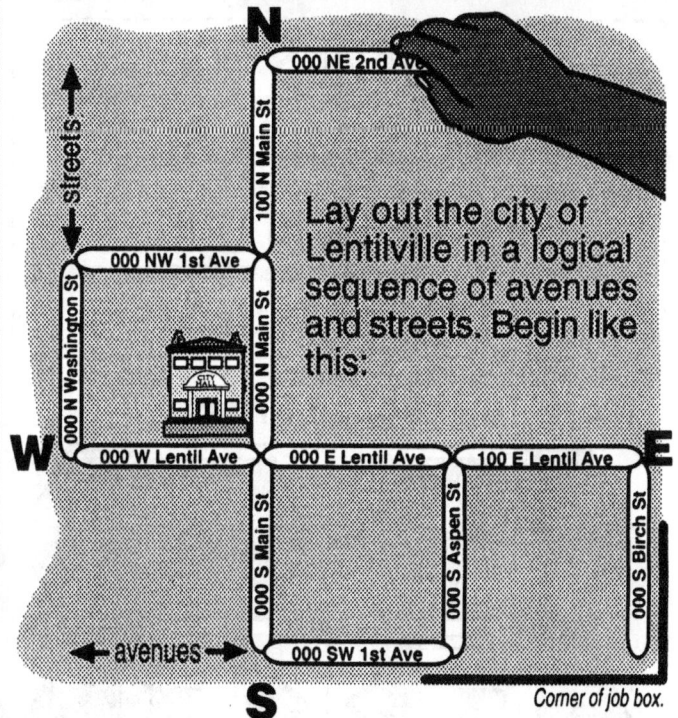

Give directions: *To get to Sam's Market from City Hall, go 2 blocks north on Main Street. Turn right at Second Avenue, and head east 1 block...*

Cut street names apart on narrow dashed lines. Fold over and glue onto craft sticks.

000 S Adams St	
000 S Adams St	
000 N Adams St	100 N Main St
000 N Adams St	100 N Main St
100 N Adams St	000 S Aspen St
100 N Adams St	000 S Aspen St
000 S Washington St	000 N Aspen St
000 S Washington St	000 N Aspen St
000 N Washington St	100 N Aspen St
000 N Washington St	100 N Aspen St
100 N Washington St	000 S Birch St
100 N Washington St	000 S Birch St
000 S Main St	000 N Birch St
000 S Main St	000 N Birch St
000 N Main St	100 N Birch St
000 N Main St	100 N Birch St

More Townscape Props design 4, 5

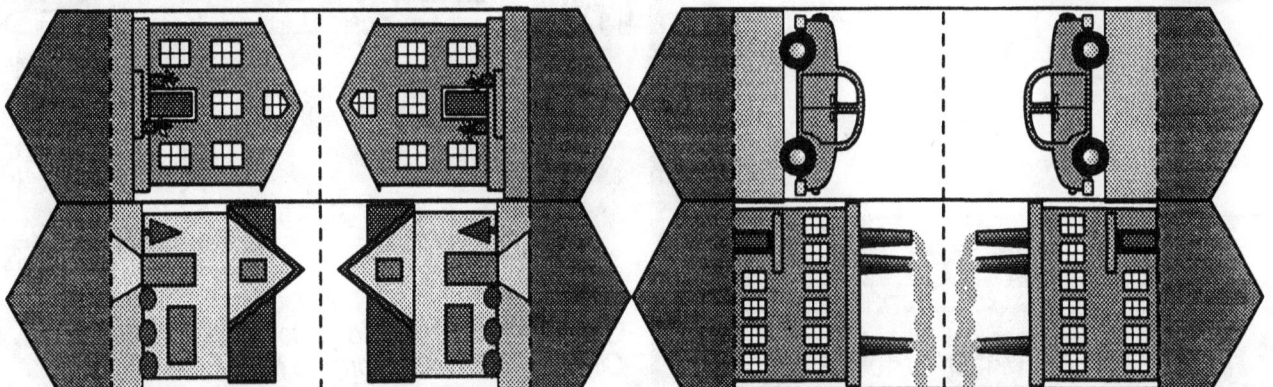

Preparation: 1 copy

100 NW 2nd Ave *(inverted)*	100 NE 2nd Ave *(inverted)*
100 NW 2nd Ave	100 NE 2nd Ave
100 NW 1st Ave *(inverted)*	100 NE 1st Ave *(inverted)*
100 NW 1st Ave	100 NE 1st Ave
100 W Lentil Ave *(inverted)*	100 E Lentil Ave *(inverted)*
100 W Lentil Ave	100 E Lentil Ave
100 SW 1st Ave *(inverted)*	100 SE 1st Ave *(inverted)*
100 SW 1st Ave	100 SE 1st Ave
000 NW 2nd Ave *(inverted)*	000 NE 2nd Ave *(inverted)*
000 NW 2nd Ave	000 NE 2nd Ave
000 NW 1st Ave *(inverted)*	000 NE 1st Ave *(inverted)*
000 NW 1st Ave	000 NE 1st Ave
000 W Lentil Ave *(inverted)*	000 E Lentil Ave *(inverted)*
000 W Lentil Ave	000 E Lentil Ave
000 SW 1st Ave *(inverted)*	000 SE 1st Ave *(inverted)*
000 SW 1st Ave	000 SE 1st Ave

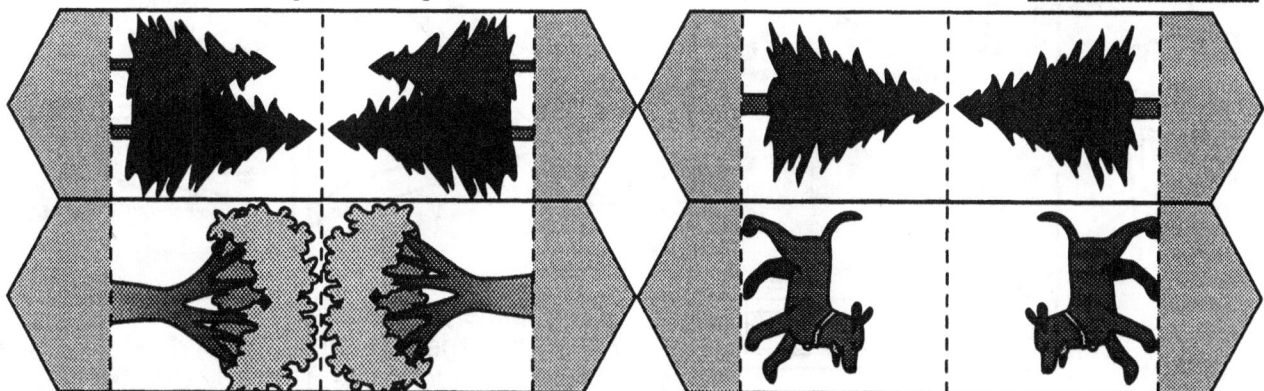

TEACHING NOTES

Purpose

To lay out Lentilville beyond the walls of the job box. To understand the logic of street addresses, and complete a city map.

Introduction

▶ Draw this block of Lentilville, in the NW section of town, on your blackboard:

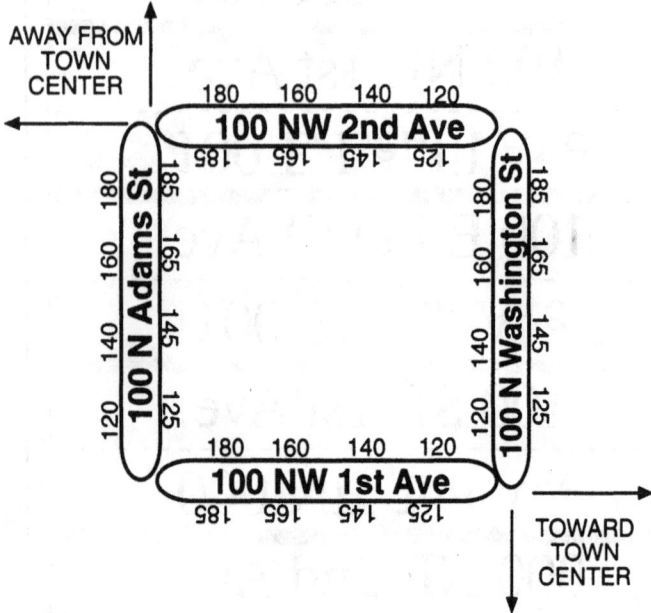

▶ All numbers along this block are "100-something." The numbers grow higher as you move farther from the center of town.

▶ Even and odd addresses appear on opposite sides of the street.

▶ Buildings facing N-S have avenue addresses. Buildings facing E-W have street addresses.

▶ Not all Street Sticks have been named yet. You may wish to follow a common American tradition, using presidents and trees. (Please use pencil, and erase the craft sticks afterward.)

• Presidents: Washington, Adams, Jefferson, Madison, Monroe, J.C. Adams, Jackson, Van Buren.

• Trees: Aspen, Birch, Cedar, Dogwood, Elm, Fir, Ginko, Holly (alphabetically).

Checkpoint

Show me your downtown map.

More

◆ Map a school bus route for picking up town children and taking them to City School. Develop a timetable for arriving at each pickup point.

◆ Plan a parade. Who will participate or contribute floats? Where will the parade begin and end? How will the police department reroute traffic during this event?

D/5 You need...

masking tape

craft sticks

street sticks

TOWNSCAPE props

job sheet

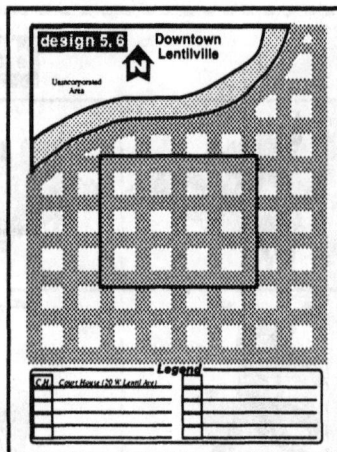

design 5

Divide your table into quadrants with masking tape. Lay out Lentilville, beginning with City Hall at 20 W. Lentil Ave.

Build outward with more craft sticks. Locate at least 10 buildings, parks or other facilities.

Complete your map. Fill in the legend with places and addresses.

Downtown Lentilville

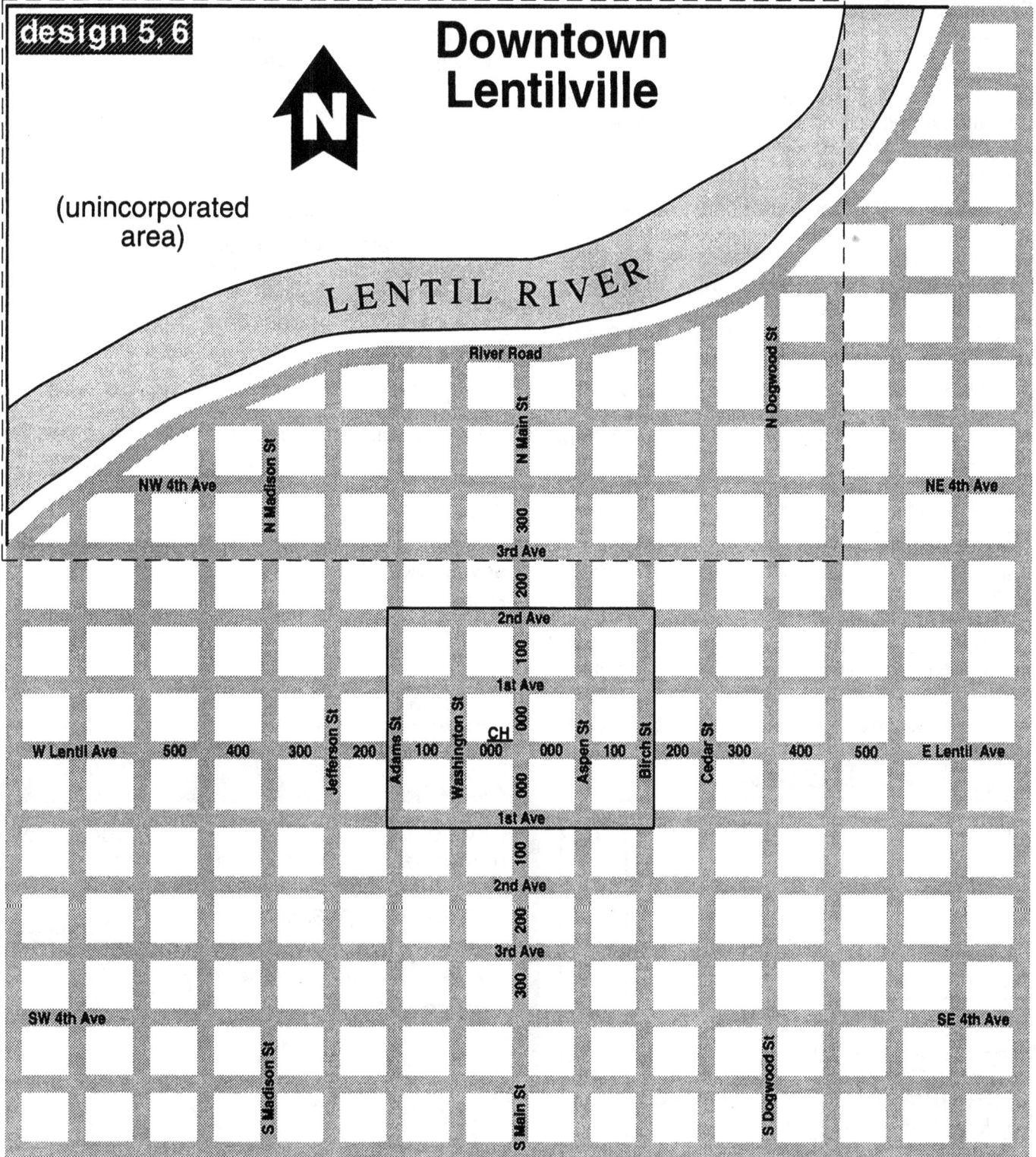

N

(unincorporated area)

LENTIL RIVER

River Road

N Madison St

N Main St

N Dogwood St

NW 4th Ave

NE 4th Ave

300

3rd Ave

200

2nd Ave

100

1st Ave

Jefferson St

Adams St

Washington St

000

CH

Aspen St

Birch St

Cedar St

W Lentil Ave | 500 | 400 | 300 | 200 | 100 | 000 | 000 | 100 | 200 | 300 | 400 | 500 | E Lentil Ave

000

1st Ave

100

2nd Ave

200

3rd Ave

300

SW 4th Ave

SE 4th Ave

S Madison St

S Main St

S Dogwood St

— Legend —

C.H.	*Court House (20 W Lentil Ave)*		

Job Sheet: 1 copy per student

TEACHING NOTES

Purpose

To plan and map an unincorporated area near Lentilville to scale. To cooperate, compromise, negotiate, balance competing interests, experience trade-offs, and problem-solve.

Introduction

◗ You will be working on a proposal to extend city boundries north of Lentil River. Please map your ideas on a new scale, 10 times larger than this downtown map. Imagine enlarging one of these tiny city blocks to look 10 times bigger. How big would that be? *A square with sides that are as long as a Street Stick.* (How convenient!)

◗ You have your own ideas about how to use the land across the river. Develop them fully, persuasively, attractively. But remember, others have their ideas, too. (After several students have completed individual proposals, you might direct them to develop and approve a final map based on class consensus.)

◗ A good plan must consider as many factors as possible. Consider any of the following that might apply to your proposal:

• How is the land now being used (farms, homes, recreation, wildlife)? Does it contain natural resources? Would people or animals be displaced by your plan? Where would they go?

• Will your plans affect air or water quality? Create noise or visual pollution for neighbors?

• Will new roads or bridges be required? Will this affect river navigation?

• What city services – water, sewer, police, fire, etc. – would be needed? What about insurance?

• How would this project be funded? Taxes? A grant? Private donations and volunteer labor? Would this project produce any income to help pay for itself?

• Who would benefit from your plan, and exactly what would those benefits be? Who might be disappointed or harmed?

Focus

What would you like to see happen across the river? Will you work alone or cooperate with others?

Checkpoint

May I see your proposal? (Are ideas well developed? Are pertinent factors considered?)

More

I notice that others have developed very different ideas. What shall we do?

D/6 You need...

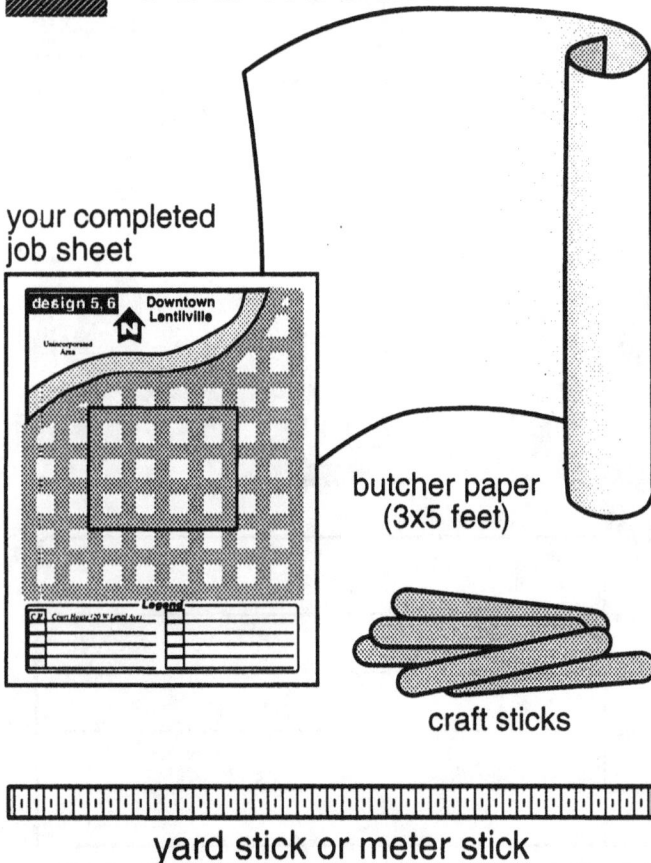

your completed job sheet

butcher paper (3x5 feet)

craft sticks

yard stick or meter stick

---

design 6 🧍 or 🧍🧍

City council is considering the purchase of unincorporated land north of Lentil River. They have directed your planning office to develop a proposal.

FARMING?
HOMES?
A PARK?
ROADS?
INDUSTRIAL?
LANDFILL?
BRIDGES?
AIRPORT?

Make each block 10 times longer than on the city map.

TEACHING NOTES
Purpose

To develop a series of drawings that represent incremental change over time. To sequence these drawings into a flip-chart movie.

Introduction

◗ Invert a liter of lentils over an empty Job Box, and let about half of them pour out. Freeze this action by capping the bottle with a lid, but continue to hold the bottle upside down.

• Ask volunteers to describe how the *bottle* looks now. How has it changed? How will it look after the remaining lentils pour out?

• Ask volunteers to describe how the *box* looks now. How has it changed? How will it look after the remaining lentils pour out?

◗ Children (and adults) often draw what they *think* they see, without really looking at the object at all. Model how to capture a true representation, not in a photographic sense, but conceptually.

• Draw where the lentils now rest in the inverted bottle. Represent the top surface with a line. Is it curved or straight? Tilted or level? Fill in with light shading and a few small circles.

• Draw the current shape of the mound in the lentil box from a side-view perspective. Is it round or pointed? Wider than it is tall?

• Add a stream of falling lentils. Watch this in action: it's the *only* detail you need to draw from memory.

Focus

◆ Which picture will you draw first? Second? Third? (Plan ahead. If you first draw the starting and ending frames, then a middle frame, you'll develop a better sense of how to pace changes in intermediate frames.)

◆ How will you represent the lentils? *With a top line and shading.* (Students who try to draw every lentil tend to draw bigger and bigger circles as they lose patience!)

Checkpoint

◆ Show me your movie. Do you see change happening at a regular pace? What would you draw differently next time?

◆ What does your picture say about gravity?

More

Draw a movie with more frames to smooth the animation. Make other movies. Maybe a large, spinning lentil (watching closely to see how light and shadows change). Draw one or more lentils falling and bouncing. Create a line of lentils wiggling like a worm. Use your imagination!

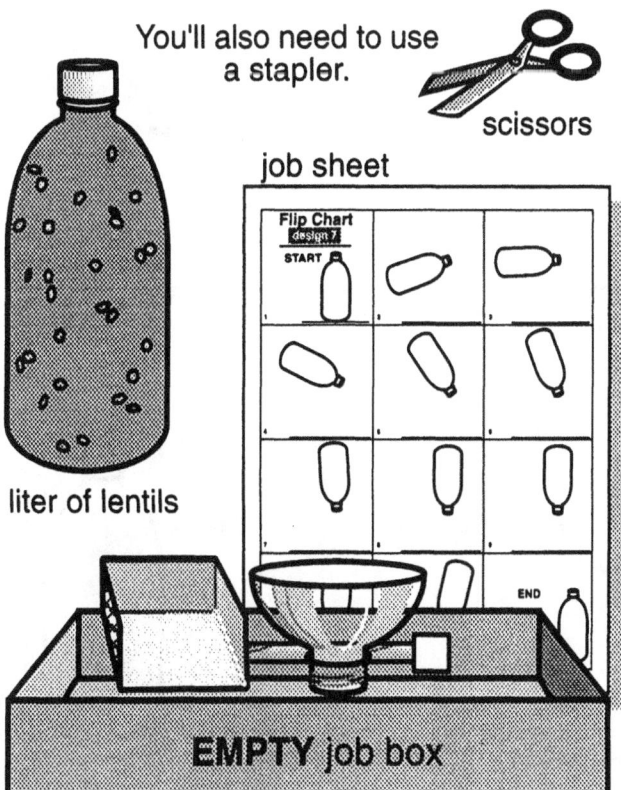

D/7 You need...

You'll also need to use a stapler.

scissors

job sheet

liter of lentils

EMPTY job box

design 7 ♀ or ♀♀

Draw 12 frames in a flip-chart movie. Plan ahead!

...frame 7...

Cut out your book. Staple the pages in order.

FAN EDGES EVENLY

STAPLE

Flip a movie!

Flip Chart
design 7

START

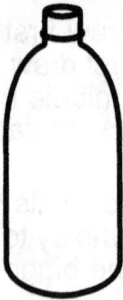

1

2

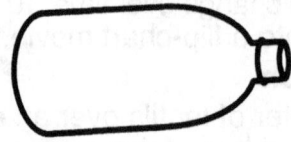

3

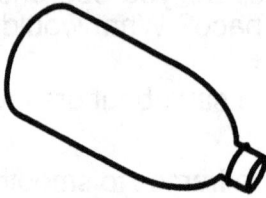

4

5

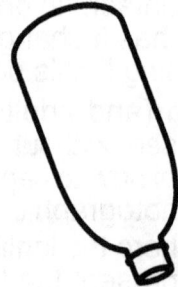

6

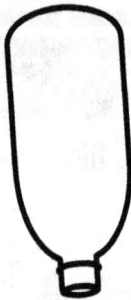

7

8

9

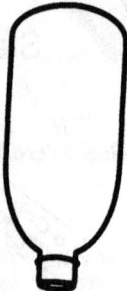

10

11

END

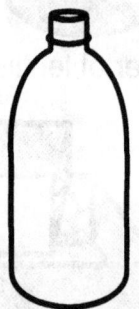

12

Job Sheet: 1 copy per student

TEACHING NOTES

Purpose

To explore the properties of an hourglass. To develop a sense of time measured in seconds.

Introduction

♦ Assemble the hourglass: Join an empty, inverted liter bottle to the top of a fair-and-full liter of loose-packed lentils. Use the Double Lid.

♦ Ask your class what they might like to investigate about this hourglass. Include the following ideas in your discussion:

• Measure time. (Establish a counting rhythm in seconds: a thousand one, a thousand two, a thousand three…, using a wall clock.)

• Experiment with slowing the hourglass down. (Insert a straight pin across the Double Lid, or measure tilt angles.)

• Study compaction. (Lentils that pour slowly through a tilted hourglass fill all of the bottom bottle, and part of the upper bottle, as well.)

• Combine different bottles. (Two liter bottles of different manufacture and shape may take longer to pour one way than the other.)

• Draw before, during and after pictures.

Checkpoint

Show me your work. What did you learn?

More

◆ Strobe effect: Computer and television screens actually blink (or refresh) at a frequency that is too fast for the human eye to see. (Our particular computer flashes 75 times per second.) If you use the light from such a monitor to backlight the falling lentils, you will see an amazing strobe effect. The flickering light captures "snapshots," "freezing" lentils in free fall, as many times per second as the brain is able to register. Instead of seeing the continuous blur illuminated by steady light, you see a much sharper "dance" of individual lentils cascading down.

◆ Static Electricity: After the upper bottle empties, numerous seeds may still cling to its sides. These lentils rubbed against plastic in their downward journey. Along the way, negatively-charged electrons rubbed off the lentils and onto the plastic. Some seeds lost so many electrons they acquired a positive charge strong enough to attract them to the electron-rich, negatively charged plastic, in defiance of earth's gravity. Watch closely; you may see individual lentils continue to slide and shift as electrons continue to move back and forth between the lentil and plastic surfaces.

D/8 You need...

double lid

straight pin

liter bottles

job box with **2 liters of lentils**

design 8   ⛊ or ⛊⛊

Make an hourglass! Experiment!

☞ *Time it.*
☞ *Slow it down.*
☞ *Study how lentils fluff & settle.*
☞ *Use different bottles.*
☞ *Draw it.*
☞ *What else can you do?*

… a thousand 5, a thousand 6…

Trial (1) = 14 seconds
Trial (2) =

TEACHING NOTES

Purpose

To provide students with an approved way to pursue their own ideas about design. To encourage creativity.

Introduction

Display this card when you want to do the activity shown here, or your own experiments.

Focus

◆ What do you want to study about design?

◆ Are these the materials you need?

Checkpoint

Report in words and pictures:
- What you did.
- What you learned.
- Questions you may still have.

Special Note

Hooray for originality! Record some of your students' most creative ideas on the opposite page.

D/9 You may need...

canning ring

tub

DESIGN
3 rocks
3 cards

job box with **1** liter of lentils

☞ Ask your teacher for other items. Tell why you need them.

---

design 9 ♀ or ♀♀

On your own.

My idea:
Lentil design puzzles...

Guess how I made these!

A | B | C | D | E | F

ANSWER KEY:

A. HANDPRINT: Duhhh.
B. RING: Rub a canning ring in a small circle.
C. BUTTERFLY: Wiggle a craft stick
D. 5-SPOKE WHEEL: Footprint of a liter bottle.
E. GALAXY: Spiral a canning ring outward.
F. BULL'S-EYE: Rub the mouth of a butter tub, then press its arched base into the center.

E / DIVIDE

In this chapter: Fractions are friendly because they make perfect pouring sense: 2/8 reduces to 1/4 because 2 eighth cups of lentils fill 1 fourth cup fair and full. Get good and comfortable solving fraction puzzles by trial and error before graduating to pencil and paper abstractions. If you fill a Prediction Tube with this part and that, do these fractions fill the whole? Lift up the tube to verify your prediction. Divide circles by degrees. Fill cups in part. Pour them together and confirm that the result adds up. Divide by height and divide by volume. Results are the same because the cups are uniform.

Basic Materials: Quantities define maximums needed to support any one Job Card in this chapter. Store *high-quantity basics* (Job Boxes, liters of lentils, bottle lids, scoops, funnels) on and under a table or counter. Store *low-quantity basics* near the "basics" sign (see page 45) or in a "basics" box. Consult our Glossary on pages 6-9 for a full description of these items. See the next page for additional special materials used in this chapter.

- [ ] **1 job box**
- [ ] **up to 2 liters of lentils**
- [ ] **1 scoop**
- [ ] **1 funnel**
- [ ] **1 craft stick**
- [ ] **extra cups**
- [ ] **masking tape**

Templates & Labels: Chapter **E / Divide**

Carefully cut out these "windshield" shapes. Wrap each around the indicated vial, flush with the bottom, and tape in place. Cut the vial to the top of the template with toenail scissors. Remove the template, cut out the label on the dashed line, and tape it on the vial.

Apply this template to a **13 or 16 dram** vial. Cut to the top of the paper.

$$\frac{1}{6}$$

sixth

Apply this template to an **8 1/2 dram** vial. Cut to the top of the paper.

$$\frac{1}{12}$$

twelveth

Apply this template to an **8 1/2 dram** vial. Cut to the top of the paper.

$$\frac{1}{10}$$

tenth

Apply this template to an **8 1/2 dram** vial. Cut to the top of the paper.

$$\frac{1}{8}$$

eighth

Apply to an **8 1/2 dram** vial. Cut to the top.

$$\frac{1}{20}$$

twentieth

Preparation: 1 copy

Store these chapter-specific items together in a designated place. They require about 1 square foot of dedicated space. General classroom materials (like scissors and tape) are also listed below when used, while others (like pencil and paper) are always assumed.

set of 10 fraction cups
(divide 1, 2, 3, 4, 5, 6, 8, 9, 10)

Store these containers in a **gallon storage jug**. You will need 4 identical sets of containers to serve a class of 30 students working simultaneously in all chapters of this curriculum. Photocopy sufficient templates and labels on pages 80 and 83.

WHOLE CUP: Use a **60 dram standard cup**. Apply the label provided.

ONE HALF: Use a **40 or 60 dram plastic vial**. Cut out the larger template that surrounds the one half label. Wrap and tape it flush with the bottom of the vial. This template now rises to the correct height. Cut to size with **toenail scissors**.

Remove the template; cut out and apply the smaller label.

ONE THIRD: Use a **20 dram plastic vial**. Cut out the template that surrounds the one third label. Wrap and tape it flush with the bottom of the vial and cut to the top of the paper.

Remove template, cut out and apply label.

FOURTH, FIFTH, SIXTH: Use **16 or 13 dram plastic vials,** as needed for height. Apply the corresponding template and cut to size.

Remove templates, cut and apply labels.

EIGHTH, TENTH, TWELFTH, TWENTIETH: Use 8$\frac{1}{2}$ **dram plastic vials**. Apply the corresponding template and cut to size.

Remove templates, cut and apply labels.

4 puzzle books (divide 2, 3)
Photocopy these **booklets** and assemble as directed.

prediction tube (divide 4)
Use any long **cardboard tube** that fits inside a standard cup. Paper towel cores, about 11 inches long, are suitable and don't need to be cut down. Or substitute two toilet tissue cores taped end to end. Cores from aluminum foil are generally made from more durable cardboard. An 11 inch length of 1$\frac{1}{4}$ inch PVC pipe is virtually indestructible.

Label with a marking pen: *prediction tube*.

blunt scissors (divide 5, 6)

circle protractor (divide 7)
Make a single photocopy of page 100. Laminate with packaging tape on both sides of the circle(s) before you cut. Reserve the rest of the page to construct a Tube Ruler (see right).

fraction labels (divide 8)
1. Place a thick rubber band around a **medium-sized can**, 3/4 of the way to the top.

2. Slip 8 **craft sticks** between the rubber band and the can. Space them evenly around the circumference. Adjust their height so they stop about a finger-width above your table, and extend about the same distance above the top of the can.

3. Tape the lower ends of the sticks to the can, using a strip of masking tape all the way around.

4. Label the tops of the craft sticks, above the rubber band: *halves, thirds, fourths, fifths, sixths, eighths, tenths and twelfths.*

5. Copy the Fraction Labels pages and trim around the outer dotted line. Cut where indicated at each arrow to separate into strips.

6. Fold each strip in half on the grey line. Tape each one closed with a long section of clear **packaging tape**, laminating them in the process.

7. Divide the strips into 6 copies each of 9 denominations of labels. Clip each denomination behind the corresponding stick on the storage can. **Paper clip** the longest strips (*twentieths*) together to store inside the can. See illustration on page 101.

measuring bottle (divide 8, 10)
Prepare the bottle illustrated on page 101 as follows:

1. Run a strip of masking tape down a liter bottle from "neck" to "foot."

2. Fill a standard cup *fair and full* with lentils; pour it into the bottle; tilt, if necessary, to level; mark the top of the lentils with a *pencil* line on the tape.

3. Calibrate in this manner up to 4 cups, maintaining a loose pack. (Do *not* settle the lentils by "tickling" or tapping the bottle. If settling does occur, invert the bottle once to "fluff," and tilt to level the contents.)

4. Label the center of each pencil line with a fine-tipped **permanent marker**: *1C, 2C, 3C, 4C*. Write "loose pack" at the top to remind students that these lines represent "fluffy" lentil volumes.

5. Ring the neck with a **rubber band**.

tube ruler (divide 9)
Cut out the rectangle along the dotted lines. Wrap it around an empty **paper towel core**, flush with one end, and trim off the excess cardboard.

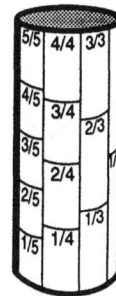

tube ruler (divide 9)

measuring bottle (divide 8, 10)

fraction labels (divide 8)

circle protractor (divide 7)

prediction tube (divide 4)

4 puzzle books (divide 2, 3)

set of 10 fraction cups
(divide 1, 2, 3, 4, 5, 6, 8, 9, 10)

E / DIVIDE

special materials

CHAPTER SIGN: Glue to a 4 x 6 inch index card, and fold in half. Stand this sign in the space where you store special materials for this chapter. Or cut this sign in half, and glue both pieces to a grocery bag that has been cut to size, or to a box. Store all listed items inside.

DIVIDE
10 containers

whole	sixth
half	eighth
third	tenth
fourth	twelfth
fifth	twentieth

DIVIDE
10 containers

whole	sixth
half	eighth
third	tenth
fourth	twelfth
fifth	twentieth

DIVIDE
10 containers

whole	sixth
half	eighth
third	tenth
fourth	twelfth
fifth	twentieth

DIVIDE
10 containers

whole	sixth
half	eighth
third	tenth
fourth	twelfth
fifth	twentieth

Apply to a gallon storage jug with clear packaging tape. (Four duplicate sets of fraction cups are recommended for large class sizes. Please duplicate pages 80 and 83 4 times for extra templates.)

Apply this template to a **40 or 60 dram** vial. Cut to the top of the paper.

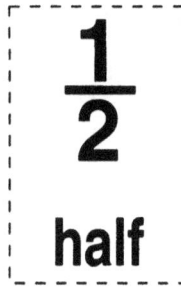

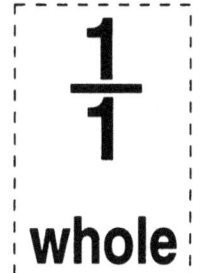

$$\frac{1}{2}$$

half

$$\frac{1}{1}$$

whole

$$\frac{1}{1}$$

whole

Apply this template to a **20 dram vial**. Cut to the top of the paper.

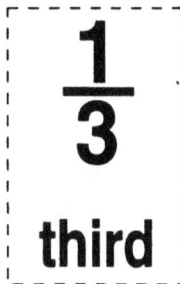

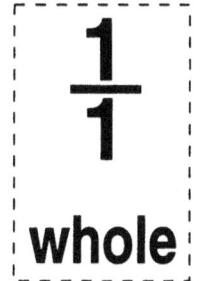

$$\frac{1}{3}$$

third

Carefully cut out these "windshield" shapes. Wrap each around the indicated vial, flush with the bottom, and tape in place. Cut the vial to the top of the template with toenail scissors. Remove the template, cut out the label on the dashed line, and tape it on the vial.

Apply this template to a **16 dram** vial. Cut to the top of the paper.

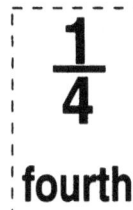

Apply this template to a **16 dram** vial. Cut to the top of the paper.

$$\frac{1}{5}$$

fifth

$$\frac{1}{4}$$

fourth

Preparation: 1 copy

LESSON NOTES

Purpose

To discover simple multiple relationships within "families" of fractions. To experience equality between reduced and unreduced fractions.

Introduction

▶ Line up the 10 fraction vials by size. Introduce them from largest to smallest: I'd like you all to meet 1 whole, 1 half, 1 third, 1 fourth, 1 fifth, 1 sixth, 1 eighth, 1 tenth, 1 twelfth, 1 twentieth.

▶ The label on each little cup tells you it is some part of the larger whole cup:

A half *is 1 of 2 equal parts in a whole;*
A third *is 1 of 3 equal parts in a whole;*
(…Keep going. Begin each sentence, then pause while your class ends each sentence in unison…)
A twentieth *is 1 of 20 equal parts in a whole;*
A hundredth *is 1 of 100 equal parts in a whole;*
A thousandth *is 1 of 1,000 equal parts…*

▶ Hold up the half: If a half is 1 of 2 equal parts in a whole, then 2 half cups should fill this whole. Demonstrate that this is true.

▶ Hold up the third. "Attempt" to verify that 3 thirds also fill the whole, but fill it about 3/4 full with each addition: 1, 2, 3, 4. Ooops! What went wrong? Let your class remind you to fill each container fair and full. (Overfill and shake once.)

▶ Write 2/8 = 1/4: These fractions are equal, though one uses bigger numbers. Who can prove it by pouring?… We can prove it with math, too, by *simplifying* each fraction. To simplify, we find a number larger than 1 that divides evenly into both numerator (top) and denominator (bottom). What number(s) will divide into 2/8? *Only the number 2.* What does that give us? *1/4.* What number will divide into 1/4? *Nothing larger than 1.* OK, then it is simplified, or reduced to its simplest terms.

Checkpoint

◆ Let me see your Job Sheet.

• Show me that your answer in this circle expresses a true relationship between these 2 vials.

• Write this relationship as an equation.

◆ Let me see your list of 18 equations. Pour lentils to show me that these two fractions are equal.

◆ Which fraction in this equation is not reduced to lowest terms? (Always the one on the left.)

◆ How do you reduce a fraction? *Divide both numerator and denominator by the same number.*

More

Keep reducing 256/1024 by twos until you can't simplify it more. Reduce 729/2187 by threes.

E/1 You need...

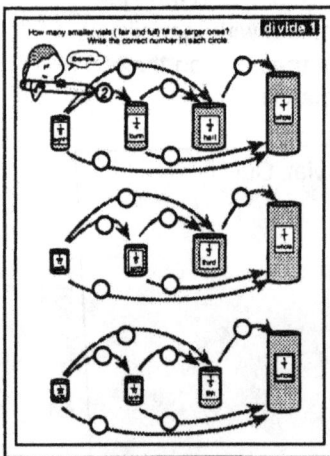

job sheet

jug of containers

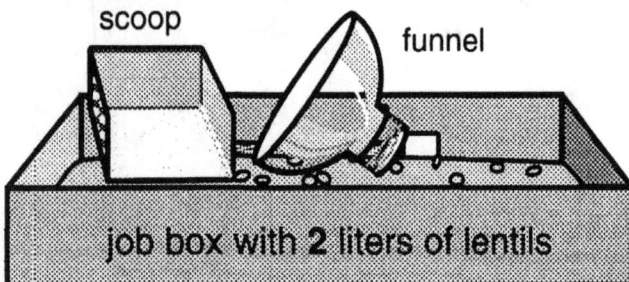

scoop   funnel

job box with **2 liters of lentils**

divide 1

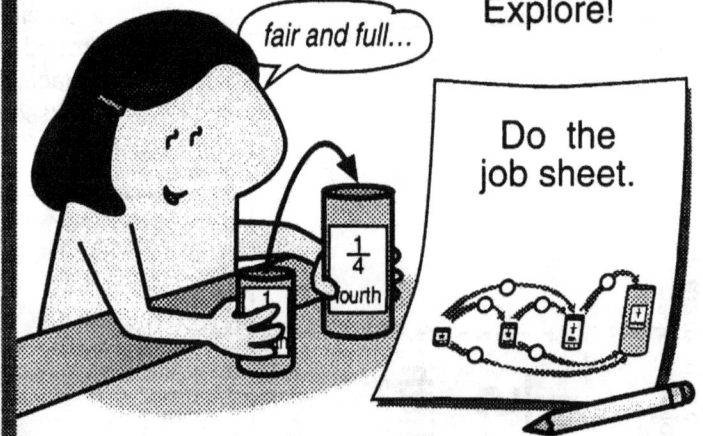

fair and full…

Explore!

Do the job sheet.

Write an equation for each circle.

EXAMPLE:

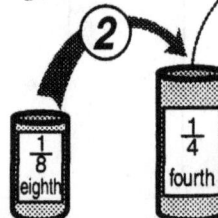

$$\frac{2}{8} = \frac{1}{4}$$

How many smaller cups (fair and full) fill the larger cups?
Write the correct number in each circle.

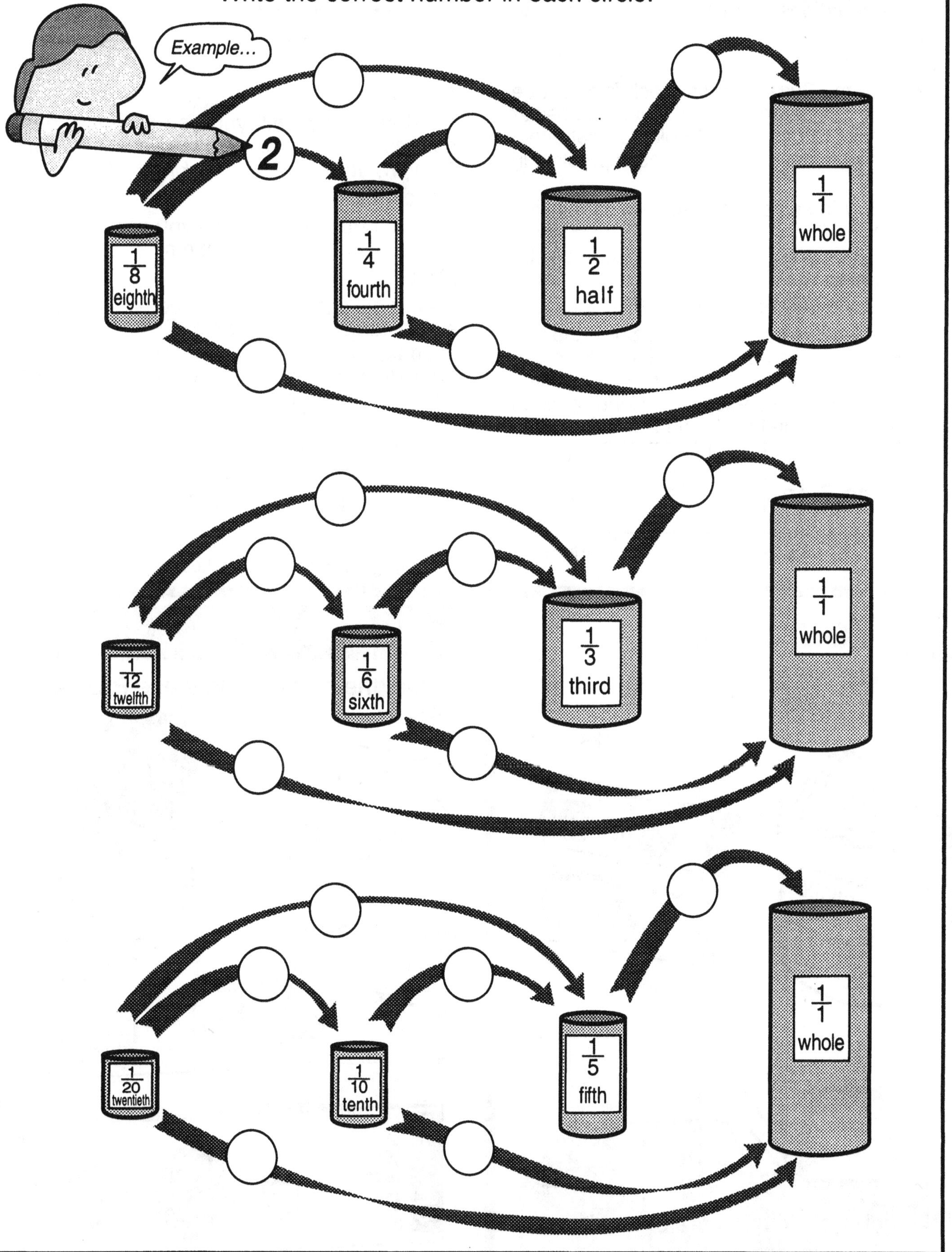

Example...

2

$\frac{1}{8}$ eighth

$\frac{1}{4}$ fourth

$\frac{1}{2}$ half

$\frac{1}{1}$ whole

$\frac{1}{12}$ twelfth

$\frac{1}{6}$ sixth

$\frac{1}{3}$ third

$\frac{1}{1}$ whole

$\frac{1}{20}$ twentieth

$\frac{1}{10}$ tenth

$\frac{1}{5}$ fifth

$\frac{1}{1}$ whole

Job Sheet: 1 copy per student

Purpose

To multiply and add fractions by pouring lentils. To solve equations by thinking with brains and hands.

Introduction

♦ Introduce the Puzzle Books. Each folded page presents a puzzle on the front, with the answer on the back. To use these books to best effect (to get smart fastest), approach each problem in 3 steps:

• First, try to solve each problem by thinking it through.

• Test your solution, or find a solution, by pouring lentils.

• Check the answer in the Puzzle Book.

♦ Make these flash card puzzles. Solve them in a 3-step process as above. The first of each pair is much easier than the second. Where brains fail, hands can easily succeed. (None of these equations appear in the Puzzle Books.)

$$\text{how many } \frac{1}{20} = \frac{1}{10}$$

2 (on back)

$$\text{how many } \frac{1}{10} = \frac{1}{4}$$

2½ (on back)

$$\frac{1}{6} + ? = \frac{1}{3}$$

$\frac{1}{6}$ (on back)

$$\frac{1}{6} + ? = \frac{1}{4}$$

$\frac{1}{12}$ (on back)

Focus

What 3-steps will you use to solve these puzzles? *Try to solve each puzzle first mentally, then physically, before checking the answer.*

Checkpoint

Which Puzzle Books did you complete? Let me see you solve this problem.

More

Write out each Puzzle Book equation on paper, substituting in the answers. Simplify (reduce) each equation to an identity. (The first problem from each book is worked as an example, with substituted answers in grey.)

$3 \times 1/6 = 1/2$ $\qquad$ $1/20 + 1/20 = 1/10$

$3/6 = 1/2$ $\qquad\qquad$ $2/20 = 1/10$

$1/2 = 1/2$ $\qquad\qquad$ $1/10 = 1/10$

E/2 You need...

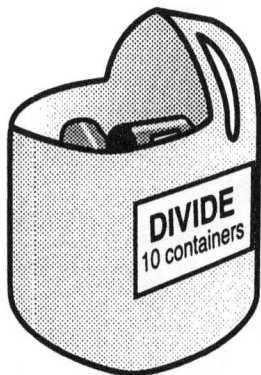

puzzle books A and B

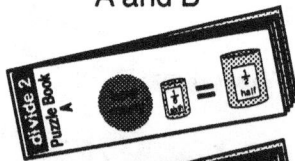

DIVIDE 10 containers

job box with **2 liters of lentils**

divide 2

Work the problems in these Puzzle Books.

*Think with your head **and** hands!*

divide 2 Puzzle Book A

$$\text{how many } \frac{1}{6} \text{ sixth} = \frac{1}{2} \text{ half}$$

divide 2 Puzzle Book B

$$\frac{1}{20} \text{ twentieth} + ? = \frac{1}{10} \text{ tenth}$$

1 | **1**

how many $\frac{1}{6}$ sixth $=$ $\frac{1}{2}$ half

3

2 | **2**

how many $\frac{1}{12}$ twelfth $=$ $\frac{1}{2}$ half

6

3 | **3**

how many $\frac{1}{10}$ tenth $=$ $\frac{1}{2}$ half

5

4 | **4**

how many $\frac{1}{20}$ twentieth $=$ $\frac{1}{2}$ half

10

5 | **5**

how many $\frac{1}{12}$ twelfth $=$ $\frac{1}{4}$ fourth

3

6 | **6**

how many $\frac{1}{20}$ twentieth $=$ $\frac{1}{4}$ fourth

5

7 | **7**

how many $\frac{1}{12}$ twelfth $=$ $\frac{1}{8}$ eighth

1$\frac{1}{2}$

8 | **8**

how many $\frac{1}{5}$ fifth $=$ $\frac{1}{2}$ half

2$\frac{1}{2}$

divide 2
puzzle book
A

Preparation: 1 copy

1 | 1

$\frac{1}{20}$ twentieth $+$ **?** $=$ $\frac{1}{10}$ tenth

$\frac{1}{20}$ twentieth

---

2 | 2

$\frac{1}{4}$ fourth $+$ $\frac{1}{4}$ fourth $+$ **?** $=$ $\frac{1}{1}$ whole

$\frac{1}{2}$ half

---

3 | 3

$\frac{1}{6}$ sixth $+$ $\frac{1}{12}$ twelfth $+$ **?** $=$ $\frac{1}{3}$ third

$\frac{1}{12}$ twelfth

---

4 | 4

$\frac{1}{20}$ twentieth $+$ $\frac{1}{20}$ twentieth $+$ **?** $=$ $\frac{1}{5}$ fifth

$\frac{1}{10}$ tenth

---

5 | 5

$\frac{1}{8}$ eighth $+$ **?** $=$ $\frac{1}{4}$ fourth

$\frac{1}{8}$ eighth

---

6 | 6

$\frac{1}{12}$ twelfth $+$ **?** $=$ $\frac{1}{6}$ sixth

$\frac{1}{12}$ twelfth

---

7 | 7

$\frac{1}{6}$ sixth $+$ **?** $=$ $\frac{1}{2}$ half

$\frac{1}{3}$ third

---

8 | 8

$\frac{1}{2}$ half $+$ $\frac{1}{3}$ third $+$ **?** $=$ $\frac{1}{1}$ whole

$\frac{1}{6}$ sixth

Preparation: 1 copy

LESSON NOTES

Purpose

To test for equalities, inequalities, and unknowns by pouring and comparing volumes. To problem-solve mentally and experimentally.

Introduction

♦ The same 3-step process presented in the previous job card applies to these Puzzle Books as well. Review as necessary.

♦ Review how to interpret inequality symbols. The *small*, pointed side of the symbol is always on the side of *less*. The *large*, open side of the symbol is always on the side of *more*. (Hungry alligators always go after the biggest meal.) Illustrate with these simple blackboard examples:

$$1 < 2 < 3 < 4$$
$$4 > 3 > 2 > 1$$

♦ Write these simple equations on the blackboard. Ask volunteers to circle the symbol that applies.

$$3 \overset{\leq}{\underset{>}{=}} 1 + 3 \qquad 4 \overset{\leq}{\underset{>}{=}} 1 + 3 \qquad 5 \overset{\leq}{\underset{>}{=}} 1 + 3$$

Focus

What 3-steps will you use to solve these puzzles? *Try to solve each puzzle first mentally, then physically, before checking the answer.*

Checkpoint

Which Puzzle Books did you complete? Let me see you solve this problem.

More

Write out each Puzzle Book equation on paper, substituting in the answers. Reduce to an identity. (The second problem from each book is worked as an example, with substituted answers in grey.)

$3 \times 1/6 > 1/3$	$1/12 + 1/12 = 1/6$
$3/6 > 1/3$	$2/12 = 1/6$
$1/2 > 1/3$	$1/6 = 1/6$

E/3 ## You need...

puzzle books A and B

job box with **2 liters of lentils**

divide 3

Do these Puzzle Books:

YAY! MORE PUZZLES!

1 $4 \times \frac{1}{8}$ eighth ? $\frac{1}{2}$ half =

2 $3 \times \frac{1}{6}$ sixth ? $\frac{1}{3}$ third >

3 $\frac{1}{8}$ eighth $+$ $\frac{1}{4}$ fourth $+$ $\frac{1}{2}$ half ? $\frac{1}{1}$ whole <

4 $2 \times \frac{1}{4}$ fourth $+$ $3 \times \frac{1}{6}$ sixth ? $\frac{1}{1}$ whole =

5 $\frac{1}{4}$ fourth $+$ $\frac{1}{5}$ fifth $+$ $\frac{1}{20}$ twentieth ? $\frac{1}{2}$ half =

6 $3 \times \frac{1}{20}$ twentieth $+$ $2 \times \frac{1}{10}$ tenth ? $\frac{1}{5}$ fifth >

7 $3 \times \frac{1}{10}$ tenth ? $\frac{1}{3}$ third < (slightly)

8 $\frac{1}{3}$ third $+$ $2 \times \frac{1}{12}$ twelfth ? $\frac{1}{2}$ half =

Preparation: 1 copy

1 | $\frac{1}{8}$ eighth + $\frac{1}{8}$ eighth = **?**   **1** | $\frac{1}{4}$ fourth

2 | $\frac{1}{12}$ + $\frac{1}{12}$ = **?**   **2** | $\frac{1}{6}$ sixth

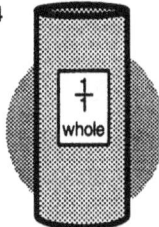

3 | $\frac{1}{20}$ + $\frac{1}{20}$ + $\frac{1}{10}$ = **?**   **3** | $\frac{1}{5}$ fifth

4 | $\frac{1}{2}$ half + $\frac{1}{4}$ fourth + $\frac{1}{4}$ fourth = **?**   **4** | $\frac{1}{1}$ whole

5 | $\frac{1}{6}$ sixth + $\frac{1}{12}$ + $\frac{1}{12}$ = **?**   **5** | $\frac{1}{3}$ third

6 | $\frac{1}{20}$ + $\frac{1}{20}$ = **?**   **6** | $\frac{1}{10}$

7 | $\frac{1}{8}$ eighth + $\frac{1}{8}$ eighth + $\frac{1}{4}$ fourth = **?**   **7** | $\frac{1}{2}$ half

8 | $\frac{1}{12}$ + $\frac{1}{4}$ fourth + $\frac{1}{6}$ sixth = **?**   **8** | $\frac{1}{2}$ half

Preparation: 1 copy

LESSON NOTES

Purpose

To review and consolidate a knowledge of fractions. To observe carefully.

Introduction

❯ Insert the Prediction Tube into an empty whole cup. Tell your class to watch closely as you add any combination of parts. Ask how full the cup will be when you pull out the Prediction Tube: will it be underfull (thumb down), fair and full (thumb level), or overfull (thumb up)? Discuss reasons. Encourage debate. Then lift the tube to reveal the answer!

Many dozens of interesting combinations are possible, at all levels of ability. Here is just a sampling.

- Prediction Tube in a *whole cup:*
    underfull: 3/8 + 2/4
    overfull: 2/12 + 2/6 + 2/3
    fair and full: 1/2 + 1/3 + 1/6
- Prediction Tube in a *half cup:*
    overfull: 3/5
    underfull: 2/5
    fair and full: 2/5 + 1/10

❯ So far, I have been pouring "fair." Is there a way of pouring that is unfair? *Yes, by NOT filling every part fair and full, so there is no logical way of predicting the end result. The game degenerates to pure guesswork.*

Focus

◆ Who will be the first pourer? The first observer?

◆ What should you do if the pourer doesn't add parts that are fair and full? *Start over.*

◆ How will you keep score? When will the game be over?

Checkpoint

◆ Watch carefully while I fill the Prediction Tube. How full is the cup / half cup?

◆ Let's play a round of predict. Who pours first?

◆ How is predicting different from guessing? *A prediction requires careful looking and thinking. A guess requires no mental effort.*

E/4 **You need...**

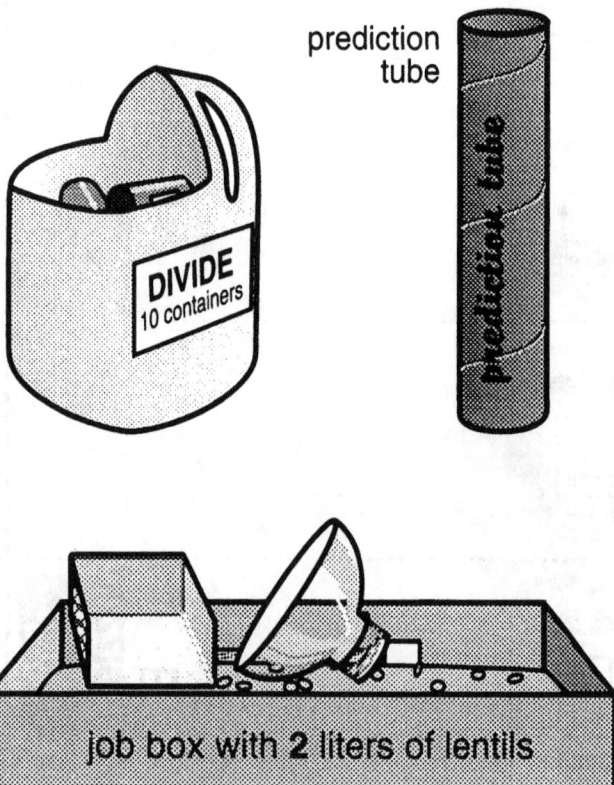

prediction tube

prediction tube

DIVIDE
10 containers

job box with **2 liters of lentils**

---

divide 4 👥

Predict right!
Win a point!

*1/2 + 1/4 + 1/5.
I think that's fair
and full.*

1/5

1/1
whole

*I think you
overfilled.*

LESSON NOTES

Purpose

To label divisions on a ruler in fractions with different denominators. To use these rulers to estimate volumes and divide lengths.

Introduction

♦ Cut out Fraction Ruler A. Fold it into a 3-strip fan.

♦ Fold the Halves section outward. Confirm that the line labeled "1/2" accurately measures the level of a half cup of lentils; that the line labeled "2/2" accurately measures the level of 2 half cups.

♦ Decide with your class how to label boxes in the fourths and eighths section of the ruler. Pour fourth cups and eighth cups into the whole to confirm that your poured fractions match the ruler marks you've labeled.

♦ Draw a narrow vertical rectangle on the blackboard the size of your fraction ruler. Use the ruler to divide this space into 4 equal parts. Shade 3 of them. This represents the fraction 3/4.

♦ Let's look closely at the "anatomy" of this fraction called 3/4. The bottom carries the *name* of the fraction (the de*nom*inator). It tells us that we are counting fourths. The top carries the *number* of the fraction (the *numer*ator). It tells us how many fourths to shade.

Focus

♦ Have you cut, folded and labeled Fraction Rulers A, B and C?

♦ Have you checked each fraction you've labeled by pouring an equal portion of lentils into a whole cup and checking its height? (Very observant students may notice that one fair-and-full portion of a very small cup, like 1/20 or 1/12, reaches higher than the corresponding line on the fraction rule. The thickness of the plastic base of the cup gives very small portions a noticeable head start.)

♦ Have you divided and shaded the Job Sheet into each given fraction?

Checkpoint

♦ Let me see Fraction Ruler C. Show me that 4/10 reaches this mark you've labeled. What other fractions reach this same level? *8/20 and 2/5.*

♦ Let me see your job sheet.

• Which part of this fraction is the numerator (number)? The denominator (name)?

• What does the denominator of each fraction tell you? *The name of the fraction. The relative size of each equal piece.*

• What does the numerator of each fraction tell you? *The number of the fraction. How many pieces to shade.*

E/5 You need...

2 job sheets

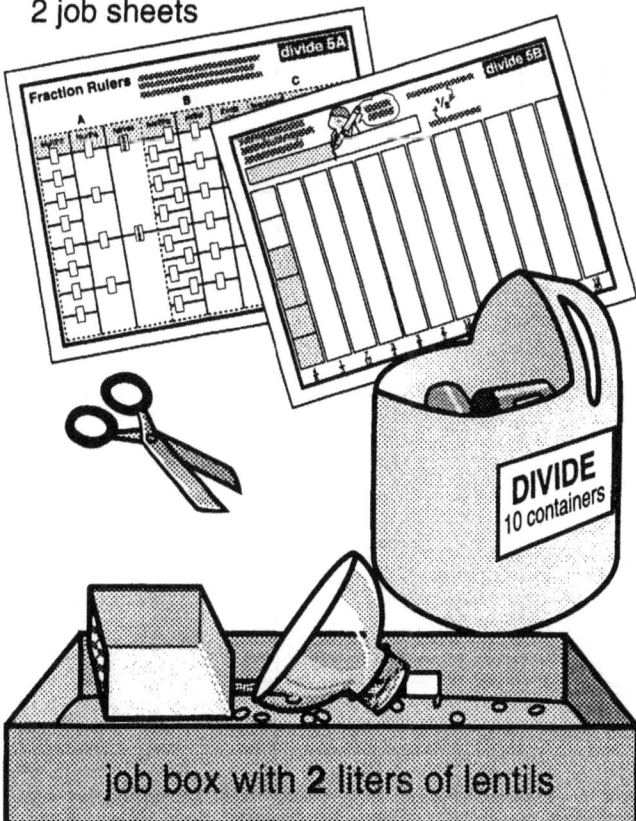

job box with **2 liters of lentils**

divide 5 ♀ or ♀♀

Cut, fold and label 3 Fraction Rulers.

Check your labels by pouring lentils.

Yes: 1/2

Use these rulers to divide and shade each bar on the job sheet.

Fraction Rulers

1. Cut out rectangles A, B and C along the dotted lines.
2. Fold each rectangle into a 3-sided fraction ruler.
3. Label each of the boxes on each ruler with the correct fraction.

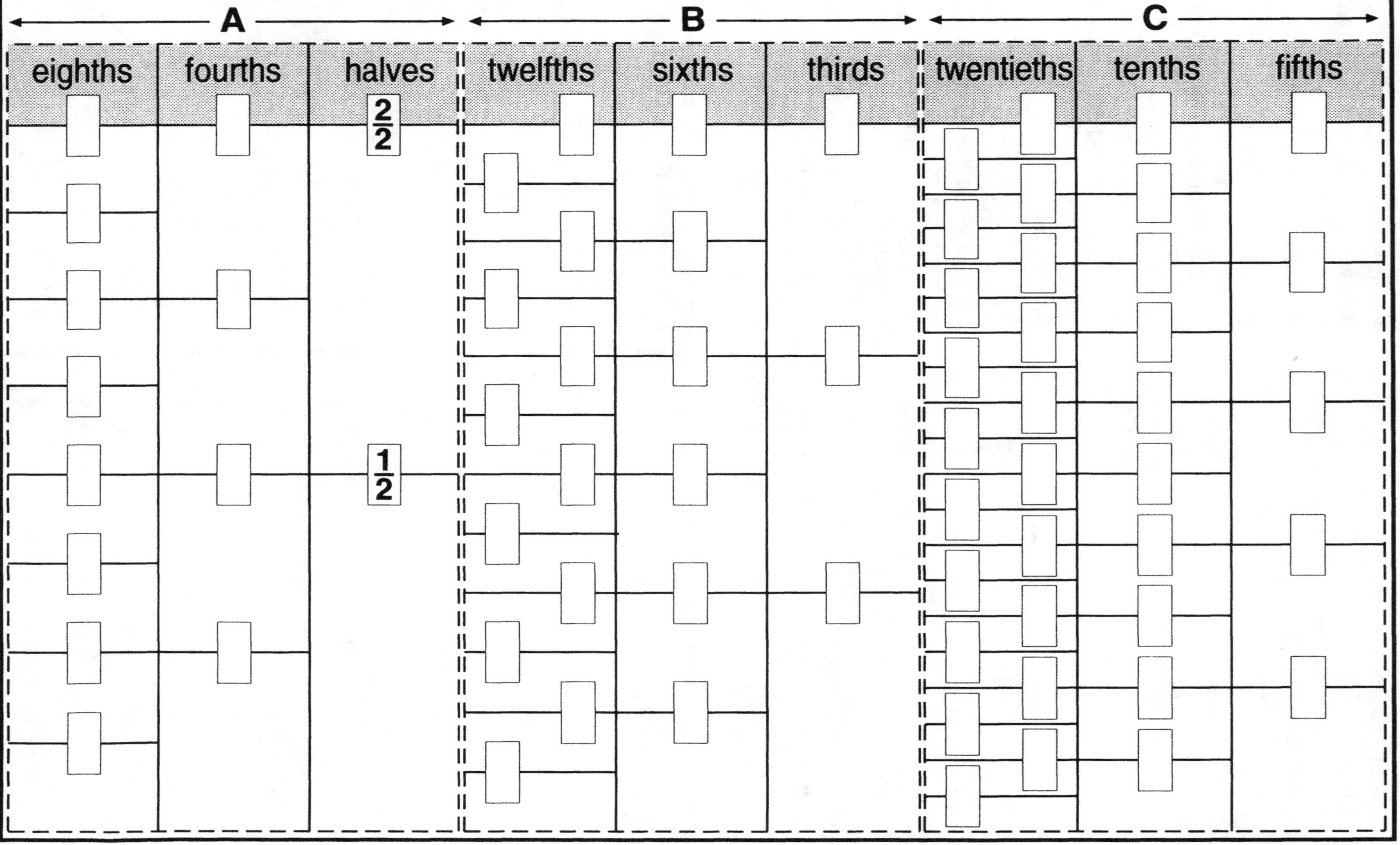

divide 5A

A

eighths

fourths

halves

$\frac{2}{2}$ $\frac{1}{2}$

B

twelfths

sixths

thirds

C

twentieths

tenths

fifths

Job Sheet: 1 copy per student

Divide each bar. Shade some of the parts.

Like this:
2 parts,
1 shaded

Divide into this many
1/2
Shade this many parts

 19/20

 1/4

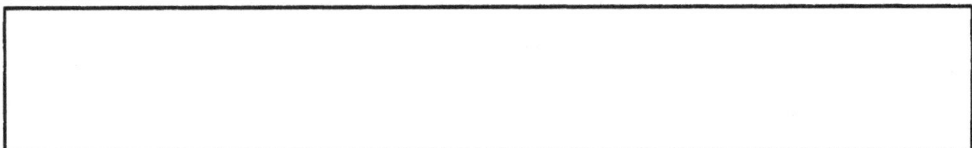 3/12

 3/5

 6/10

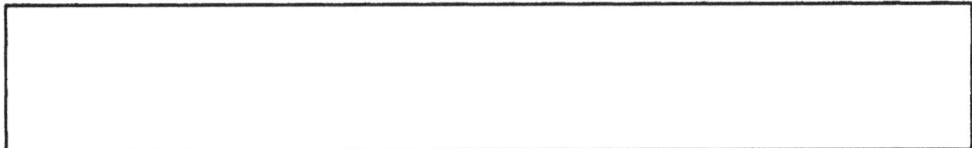 12/20

 2/2

 3/8

 3/4

 7/12

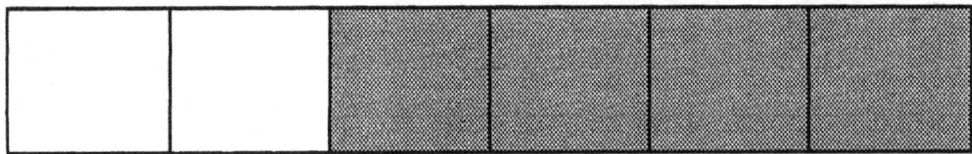 1/3

4/6

Job Sheet: 1 copy per student

LESSON NOTES

Purpose

To pour fractional portions of lentils into a whole cup, and record the levels. To notice patterns.

Introduction

◗ Cut out the Multi-Strip with the *solid* fold lines. Fold it forward and back along each line, creasing it both ways.

◗ Read across the top of the strip. We must partly fill a whole with each of these fractional parts, one at a time.

• Let's start at the beginning, with 1/1. Fold that face of the Multi-Strip to the outside, so we can stand it beside an empty whole cup. Pour the 1/1 cup, fair and full, into the empty cup. Mark the lentil level with a pencil line and shade the area underneath.

• Fill and shade the level for 1/2 cup in the same manner. Can anyone guess how the completed Multi-Strip will look? (Accept all predictions, but don't discuss. Allow students to discover patterns as they complete the Job Card.)

Focus

◆ Which Multi-Strips should you cut and shade first? *The ones with solid lines.*

◆ Can you discover any patterns?

Checkpoint

◆ Show me the *solid*-line Multi-Strip:

• What pattern do you see? *A descending staircase, with smaller and smaller steps.*

• Why do we get this pattern? *We keep measuring just 1 part that forms a smaller and smaller fraction of the whole.*

• What happens to fractions as their denominators get very large? *Their size (or stairsteps) get very small, approaching zero.*

◆ Show me the *dotted*-line Multi-Strip:

• What pattern do you see? *A valley with a level floor.*

• Why do we get this pattern? *From left to right, each step downward is 1/4 less than the step before. Then a few stips flatten out: these all measure fractions equal to 1/4. Then larger and larger fractions step upward again.*

More

Check the accuracy of each Multi-Strip with your Fraction Ruler. Do you get the same result? (Approximately. The thickness of the plastic base of the cup gives very small portions a higher level than corresponding divisions on the Fraction Ruler.)

E/6 You need...

DIVIDE
10 containers

job sheet

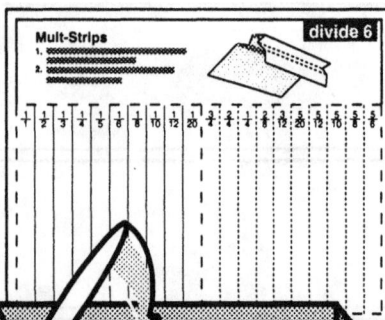

job box with 2 liters of lentils

divide 6

Cut and fold the Multi-Strip with *solid* lines.

Mark and shade how high each fraction cup fills the whole cup.

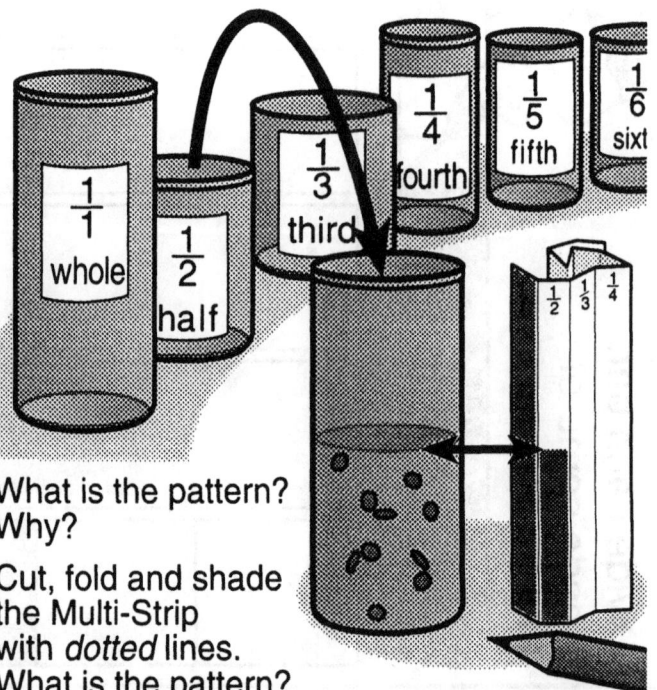

What is the pattern? Why?

Cut, fold and shade the Multi-Strip with *dotted* lines. What is the pattern?

divide 6

Multi-Strips

1. Cut out both the **solid** and **dotted** rectangles along the bold dashed lines.
2. Fold and crease forward and backward along the solid and dotted lines.

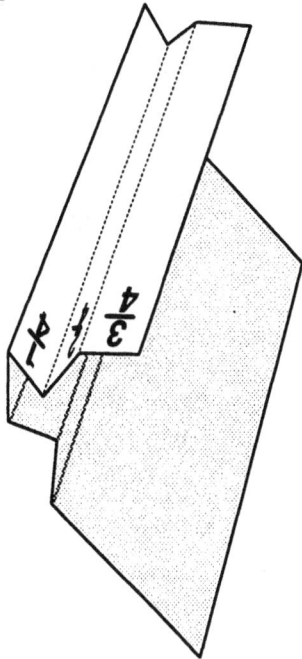

$\frac{5}{6}$	$\frac{5}{8}$	$\frac{5}{10}$	$\frac{5}{12}$	$\frac{5}{20}$	$\frac{3}{12}$	$\frac{2}{8}$	$\frac{1}{4}$	$\frac{2}{4}$	$\frac{3}{4}$

| $\frac{1}{20}$ |
| $\frac{1}{12}$ |
| $\frac{1}{10}$ |
| $\frac{1}{8}$ |
| $\frac{1}{6}$ |
| $\frac{1}{5}$ |
| $\frac{1}{4}$ |
| $\frac{1}{3}$ |
| $\frac{1}{2}$ |
| $\frac{1}{1}$ |

Job Sheet: 1 copy per student

LESSON NOTES

Purpose

To represent fractions by dividing circles with a protractor, and shading some of the parts. To extend the concept of fractions to area.

Introduction

◗ Sketch a large Circle Protractor on your blackboard. Keep all calibrations and numbers to the *outside,* as illustrated, so you can draw and erase inside division lines without erasing the protractor.

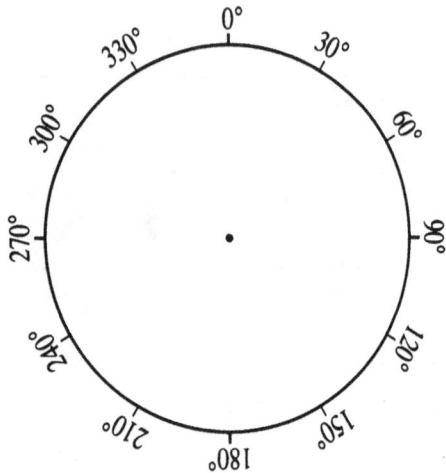

◗ Pretend this protractor is a giant pie to be shared equally among 2, 3 and 12 people. Divide these numbers into 360° to determine where you will cut each "pie."

Focus

◆ How will you know where to divide each pie? (Some fractions are easy. But students will have to divide denominators into 360° for less intuitive divisions.)

◆ Where will you show your math? *On the front or back of the Job Sheet.*

Checkpoint

◆ What does the denominator of a fraction tell you? *Its name. The total number of pieces.*

◆ What does the numerator of a fraction tell you? *How many pieces to shade.*

◆ Show me your Job Sheet. What calculations led you to draw this line here? (Once the number of degrees between one division is determined, the pie can be divided by simply rotating the protractor. Consider keeping this trick to yourself so students don't miss good multiplication practice. Also, the divisions tend to be less accurate.)

E/7 You need...

job sheet

craft stick

circle protractor

divide 7 ⭢ or ⭢⭢

Divide and shade the correct part of each pie.

Use the Circle Protractor.

Show your math.

$$4\overline{)360°}$$

divide 7

Use the Circle Protractor to divide each circle into equal parts. Lightly shade the correct number.

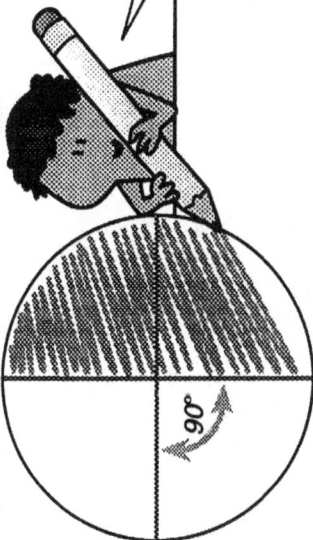

Divide into this many

2/4

Shade this many parts

90°

1/2

3/4

5/8

2/3

2/4

5/6

3/10

11/12

3/20

1/36

Circle Protractor

 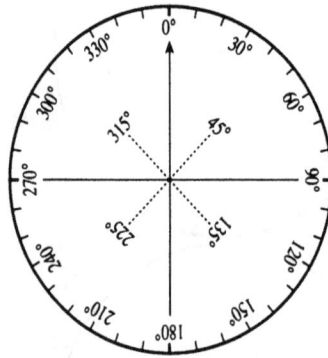 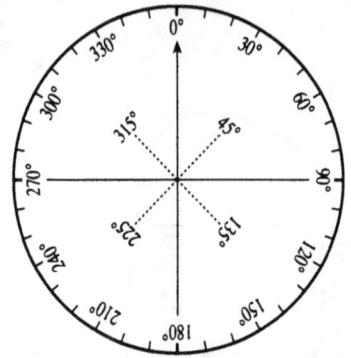

Tube Ruler

20/20	12/12	10/10	8/8	6/6	5/5	4/4	3/3	2/2
19/20								
18/20	11/12	9/10						
17/20			7/8					
16/20	10/12	8/10		5/6				
15/20	9/12		6/8		4/5	3/4		
14/20		7/10						
13/20	8/12			4/6			2/3	
12/20		6/10	5/8		3/5			
11/20	7/12							
10/20	6/12	5/10	4/8	3/6		2/4		1/2
9/20								
8/20	5/12	4/10			2/5			
7/20			3/8					
6/20	4/12	3/10		2/6			1/3	
5/20	3/12		2/8			1/4		
4/20		2/10			1/5			
3/20	2/12			1/6				
2/20		1/10	1/8					
1/20	1/12							

Preparation: 1 copy

LESSON NOTES

Purpose

To create fractions, predict sums, verify by pouring them together, and record the resulting equations.

Introduction

◆ Hold up the can that organizes the Fraction Labels: Notice that there are 9 denominations of labels, 6 copies each, written on both sides.

◆ Select the thirds. All these fractions have "3" as the denominator. Place the grey patch beneath "2/3" behind the paper clip on a whole cup. Invite a volunteer to read this fraction, then pour in the thirds cup twice to fill the whole cup 2/3 full.

◆ Label and fill 3 more cups 1/3 or 2/3 full, as pictured on the Job Card. Discuss how high all these lentils might reach when poured together into the measuring bottle. Then position a rubber band over the predicted level. (The "peoplet" predicts 2 cups. Other predictions are fine, too. Just add more rubber bands.)

◆ Funnel each container into the Measuring Bottle. Point out that you have avoided "tickling" or shaking the bottle, since it is calibrated for loose fill. Summarize the result with an equation:

$$1/3 + 1/3 + 2/3 + 2/3 = 6/3 = 2$$

◆ Practice predicting levels that fall *between* whole cup calibrations. Discuss where these sums fall between the 1 and 2 cup lines. Confirm by pouring: (1/1 + 1/2), (1/1 + 1/3), (1/1 + 2/3), (1/1 + 1/4), (1/1 + 3/4), (1/1 + 4/5), (1/1 + 2/5), (1/1 + 1/6).

Focus

◆ Which denomination(s) will you use? (Suggestions in order of difficulty:

Single denominators: halves, fourths, thirds, eighths, fifths, tenths, sixths, twentieths, twelfths.

Common denominators: halves/fourths/eights; fifths/tenths/twentieths; thirds/sixths/twelfths.

Mixed denominators.)

◆ How many whole cups will you use? *Up to 6.* (This allows students to generate sums with as few as 2 terms, to as many as 6.)

Checkpoint

◆ May I see your equations? Pour lentils to prove that the sum you have written here is correct.

◆ I notice you have been using fractions like these. Can you generate sums using these other combinations? (Gently nudge students a little beyond their comfort zones, where more learning can happen. These manipulatives can challenge students from 6 to 60.)

E/8 You need...

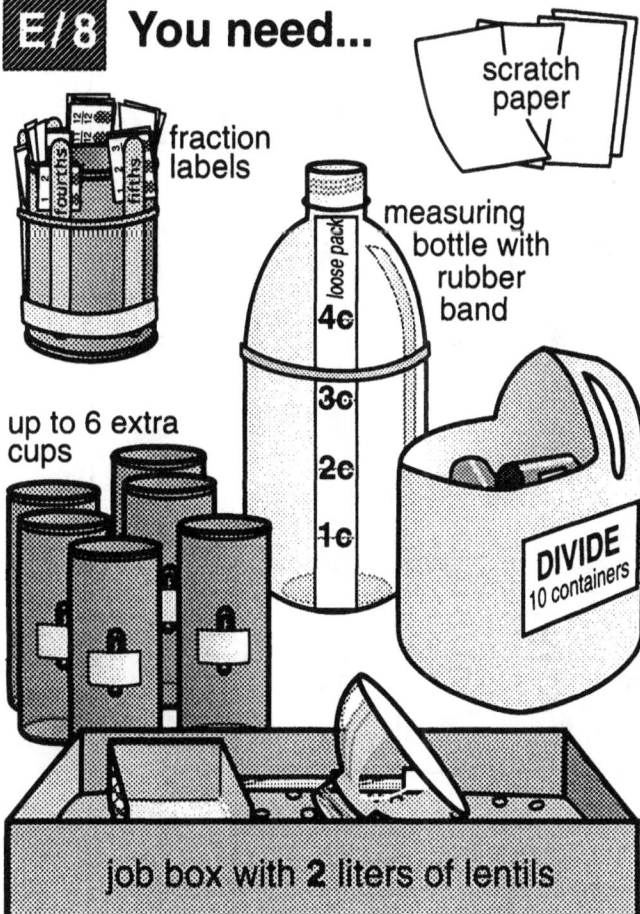

scratch paper

fraction labels

measuring bottle with rubber band

up to 6 extra cups

DIVIDE 10 containers

job box with **2 liters of lentils**

---

divide 8 　♀ or ♀♀

Partly fill cups, and label.

Predict how much.

I think it will total 2 cups.

Test. Write the equation.

Fraction Labels

Copyright © 1999 by TOPS Learning Systems

Preparation: 1 copy

$\frac{1}{8}$	$\frac{2}{8}$	$\frac{3}{8}$	$\frac{4}{8}$	$\frac{1}{20}$	$\frac{2}{20}$	$\frac{3}{20}$	$\frac{4}{20}$	$\frac{5}{20}$	$\frac{6}{20}$	$\frac{7}{20}$	$\frac{8}{20}$	$\frac{9}{20}$	$\frac{10}{20}$
$\frac{8}{8}$	$\frac{7}{8}$	$\frac{6}{8}$	$\frac{5}{8}$	$\frac{20}{20}$	$\frac{19}{20}$	$\frac{18}{20}$	$\frac{17}{20}$	$\frac{16}{20}$	$\frac{15}{20}$	$\frac{14}{20}$	$\frac{13}{20}$	$\frac{12}{20}$	$\frac{11}{20}$

$\frac{1}{8}$	$\frac{2}{8}$	$\frac{3}{8}$	$\frac{4}{8}$	$\frac{1}{20}$	$\frac{2}{20}$	$\frac{3}{20}$	$\frac{4}{20}$	$\frac{5}{20}$	$\frac{6}{20}$	$\frac{7}{20}$	$\frac{8}{20}$	$\frac{9}{20}$	$\frac{10}{20}$
$\frac{8}{8}$	$\frac{7}{8}$	$\frac{6}{8}$	$\frac{5}{8}$	$\frac{20}{20}$	$\frac{19}{20}$	$\frac{18}{20}$	$\frac{17}{20}$	$\frac{16}{20}$	$\frac{15}{20}$	$\frac{14}{20}$	$\frac{13}{20}$	$\frac{12}{20}$	$\frac{11}{20}$

$\frac{1}{8}$	$\frac{2}{8}$	$\frac{3}{8}$	$\frac{4}{8}$	$\frac{1}{20}$	$\frac{2}{20}$	$\frac{3}{20}$	$\frac{4}{20}$	$\frac{5}{20}$	$\frac{6}{20}$	$\frac{7}{20}$	$\frac{8}{20}$	$\frac{9}{20}$	$\frac{10}{20}$
$\frac{8}{8}$	$\frac{7}{8}$	$\frac{6}{8}$	$\frac{5}{8}$	$\frac{20}{20}$	$\frac{19}{20}$	$\frac{18}{20}$	$\frac{17}{20}$	$\frac{16}{20}$	$\frac{15}{20}$	$\frac{14}{20}$	$\frac{13}{20}$	$\frac{12}{20}$	$\frac{11}{20}$

$\frac{1}{8}$	$\frac{2}{8}$	$\frac{3}{8}$	$\frac{4}{8}$	$\frac{1}{20}$	$\frac{2}{20}$	$\frac{3}{20}$	$\frac{4}{20}$	$\frac{5}{20}$	$\frac{6}{20}$	$\frac{7}{20}$	$\frac{8}{20}$	$\frac{9}{20}$	$\frac{10}{20}$
$\frac{8}{8}$	$\frac{7}{8}$	$\frac{6}{8}$	$\frac{5}{8}$	$\frac{20}{20}$	$\frac{19}{20}$	$\frac{18}{20}$	$\frac{17}{20}$	$\frac{16}{20}$	$\frac{15}{20}$	$\frac{14}{20}$	$\frac{13}{20}$	$\frac{12}{20}$	$\frac{11}{20}$

$\frac{1}{8}$	$\frac{2}{8}$	$\frac{3}{8}$	$\frac{4}{8}$	$\frac{1}{20}$	$\frac{2}{20}$	$\frac{3}{20}$	$\frac{4}{20}$	$\frac{5}{20}$	$\frac{6}{20}$	$\frac{7}{20}$	$\frac{8}{20}$	$\frac{9}{20}$	$\frac{10}{20}$
$\frac{8}{8}$	$\frac{7}{8}$	$\frac{6}{8}$	$\frac{5}{8}$	$\frac{20}{20}$	$\frac{19}{20}$	$\frac{18}{20}$	$\frac{17}{20}$	$\frac{16}{20}$	$\frac{15}{20}$	$\frac{14}{20}$	$\frac{13}{20}$	$\frac{12}{20}$	$\frac{11}{20}$

CUT

$\frac{1}{12}$	$\frac{2}{12}$	$\frac{3}{12}$	$\frac{4}{12}$	$\frac{5}{12}$	$\frac{6}{12}$	$\frac{1}{10}$	$\frac{2}{10}$	$\frac{3}{10}$	$\frac{4}{10}$	$\frac{5}{10}$
$\frac{12}{12}$	$\frac{11}{12}$	$\frac{10}{12}$	$\frac{9}{12}$	$\frac{8}{12}$	$\frac{7}{12}$	$\frac{10}{10}$	$\frac{9}{10}$	$\frac{8}{10}$	$\frac{7}{10}$	$\frac{6}{10}$

$\frac{1}{12}$	$\frac{2}{12}$	$\frac{3}{12}$	$\frac{4}{12}$	$\frac{5}{12}$	$\frac{6}{12}$	$\frac{1}{10}$	$\frac{2}{10}$	$\frac{3}{10}$	$\frac{4}{10}$	$\frac{5}{10}$
$\frac{12}{12}$	$\frac{11}{12}$	$\frac{10}{12}$	$\frac{9}{12}$	$\frac{8}{12}$	$\frac{7}{12}$	$\frac{10}{10}$	$\frac{9}{10}$	$\frac{8}{10}$	$\frac{7}{10}$	$\frac{6}{10}$

$\frac{1}{12}$	$\frac{2}{12}$	$\frac{3}{12}$	$\frac{4}{12}$	$\frac{5}{12}$	$\frac{6}{12}$	$\frac{1}{10}$	$\frac{2}{10}$	$\frac{3}{10}$	$\frac{4}{10}$	$\frac{5}{10}$
$\frac{12}{12}$	$\frac{11}{12}$	$\frac{10}{12}$	$\frac{9}{12}$	$\frac{8}{12}$	$\frac{7}{12}$	$\frac{10}{10}$	$\frac{9}{10}$	$\frac{8}{10}$	$\frac{7}{10}$	$\frac{6}{10}$

$\frac{1}{12}$	$\frac{2}{12}$	$\frac{3}{12}$	$\frac{4}{12}$	$\frac{5}{12}$	$\frac{6}{12}$	$\frac{1}{10}$	$\frac{2}{10}$	$\frac{3}{10}$	$\frac{4}{10}$	$\frac{5}{10}$
$\frac{12}{12}$	$\frac{11}{12}$	$\frac{10}{12}$	$\frac{9}{12}$	$\frac{8}{12}$	$\frac{7}{12}$	$\frac{10}{10}$	$\frac{9}{10}$	$\frac{8}{10}$	$\frac{7}{10}$	$\frac{6}{10}$

$\frac{1}{12}$	$\frac{2}{12}$	$\frac{3}{12}$	$\frac{4}{12}$	$\frac{5}{12}$	$\frac{6}{12}$	$\frac{1}{10}$	$\frac{2}{10}$	$\frac{3}{10}$	$\frac{4}{10}$	$\frac{5}{10}$
$\frac{12}{12}$	$\frac{11}{12}$	$\frac{10}{12}$	$\frac{9}{12}$	$\frac{8}{12}$	$\frac{7}{12}$	$\frac{10}{10}$	$\frac{9}{10}$	$\frac{8}{10}$	$\frac{7}{10}$	$\frac{6}{10}$

CUT

Preparation: 1 copy

LESSON NOTES

Purpose

To measure fractional portions of lentils in a whole cup by height and by volume.

Introduction

▶ Scoop any unmeasured portion of lentils into the whole cup. Estimate its volume *by height* using the Tube Ruler:

• Stand the tube on a level surface beside the partially filled cup.

• Rotate the tube to the find the line that best matches the level of lentils in the cup. (Suppose it is closest to the 7/8 mark.)

• Express this measure in one of 3 ways: 7/8⁺ if you think the cup contains a little more; 7/8° if you think the cup contains precisely that much; 7/8⁻ if you think the cup contains a little less.

▶ Measure this same portion *by volume* using the appropriate fraction cup. See if the estimated volume of 7/8 fills the eighths cup 7 times.

• Use your hand to help funnel and control the lentils as they flow into the smaller container. (The large funnel is too prone to overfill.)

• Express your answer in one of 3 ways as before: 7/8⁺, 7/8° or 7/8⁻. Height measure and volume measure will seldom agree exactly.

Focus

◆ Tell me 2 ways you can measure lentils. *By height, using the Tube Ruler; by volume, using measuring cups.*

◆ Can you organize both measurements in a table? Use +, o, and - to express small differences.

Checkpoint

◆ Show me your table of height measure and volume measure. Do figures in the two columns always match? *Not exactly.* (Measurement carries an unavoidable degree of uncertainty. Lentils can fluff or settle. Empty, for the cup, is a bit higher than zero on the Ruler because of its thick bottom.)

◆ Fill this whole cup with any unmeasured portion of lentils. Measure it by height and by volume.

More

Play this cooperative guessing game:

• Take turns filling the whole cup with a *measured* fraction of lentils while your friend looks away. Write this number on scratch paper.

• Your friend then uses the Tube Ruler to guess what you added. Score 1 point for each correct answer. (Small differences of +, o, and - are OK.) Subtract 2 points for each incorrect answer. If you both reach 5, you both win!

E/9 You need...

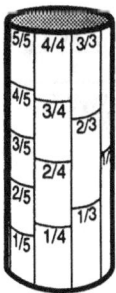

tube ruler

DIVIDE 10 containers

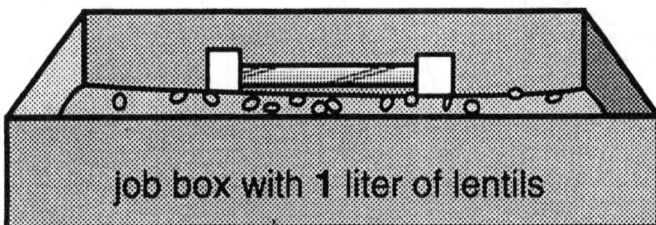

job box with 1 liter of lentils

---

divide 9

Fill a whole cup with any amount of lentils. Measure in 2 different ways:

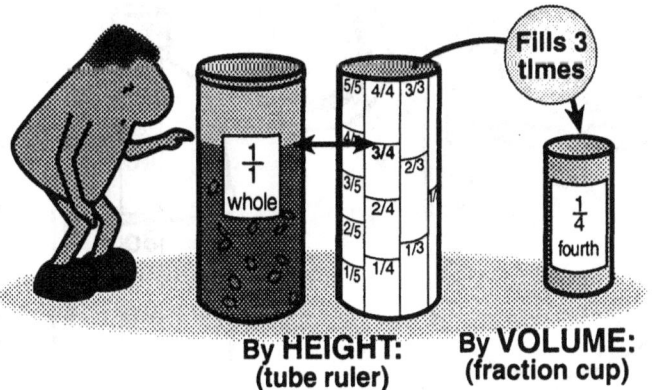

Fills 3 times

By **HEIGHT:** (tube ruler)

By **VOLUME:** (fraction cup)

Record your results in a table.

tube	cup
7/8°	7/8⁻
3⁺/5	3/5

Record all the *equivalent* fractions you can find on the Tube Ruler.

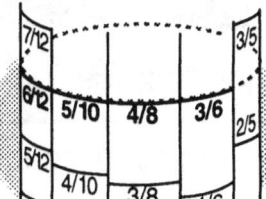

TEACHING NOTES

Purpose
To provide students with an approved way to pursue their own ideas about dividing.

Introduction
Display this card when you want to do the suggested activity or your own experiment.

Focus
◆ What do you want to study about dividing?
◆ Are these the materials you need?

Checkpoint
Report in words and pictures:
- What you did.
- What you learned.
- Questions you may still have.

Special Note
Stimulate new ideas: List the best ones on the opposite page!

E/10 You need...

job box with up to **2 liters of lentils**

☞ Ask your teacher for other items.
Tell why you need them.

DIVIDE
10 containers

loose pack
4c
3c
2c
1c

---

divide 10   ☖ or ☖☖

On your own.

My idea:
Use masking tape to label all fraction vials with a common denominator of 120.

Write equations two ways:
Unreduced:
  $20/120 + 40/120 = 60/120$
Reduced: $1/6 + 1/3 = 1/2$

third third

$\frac{60}{120}$ half

$\frac{1}{1}$ $\frac{120}{120}$ whole

F / CALIBRATE

In this chapter: Divide a liter into equal parts. Pour each part back into the bottle to calibrate in fractions. Calibrate a half-liter bottle in the same manner. Pour half-liter measure into whole-liter measure to understand in concrete terms why half of a half equals a quarter. Graph how height changes with volume on a line graph. Correlate the slope of the graph line with the size of the containers you use. Notice how the shape of a container makes the graph line turn up or down. Organize data in tables and plot ordered pairs.

Basic Materials: Quantities define maximums needed to support any one Job Card in this chapter. Store *high-quantity basics* (Job Boxes, liters of lentils, bottle lids, scoops, funnels) on and under a table or counter. Store *low-quantity basics* near the "basics" sign (see page 45) or in a "basics" box. Consult our Glossary on pages 6-9 for a full description of these items. See the next page for additional special materials used in this chapter.

- [ ] **1 job box**
- [ ] **up to 2 liters of lentils**
- [ ] **3 tubs**
- [ ] **8 clear cups**
- [ ] **1 funnel**
- [ ] **1 scoop**
- [ ] **clear tape**
- [ ] **extra cups**
- [ ] **masking tape**

Store these chapter-specific items together in a designated place. They require about 1/2 square foot of dedicated space. General classroom materials (like scissors and tape) are also listed below when used, while others (like pencil and paper) are always assumed.

blunt scissors (calibrate 1, 2, 3, 4, 5)

half-liter bottle (calibrate 2, 5, 7)

Find this in the grocery store or convenience store cold case. Best shape is tall and straight. Ribbed sides are OK.

colored pencils, crayons or markers (calibrate 3)

This is usually a student supply.

ruler or straight edge (calibrate 4, 5, 6)

This is usually a student supply

baby food jars (calibrate 4, 5, 7)

Provide small, medium and large sizes.

ruled cups (calibrate 6)

Use a variety of containers, approximately cup-sized, with a diversity of shape. We recommend using a dedicated **standard cup**, a **clear cup** and a **250 ml Erlenmeyer flask**. (You might borrow this flask from a school lab.) Photocopy the centimeter rulers on the next page. Fix one each to the side of each container with clear packaging tape, so the zero mark is flush with the bottom. Trim at the top.

film can (calibrate 6)

Use a film canister, or cut a small plastic bottle to approximately the same volume.

set of 10 fraction cups (calibrate 7)

Borrow this from Divide.

Labels for Basic Materials

EXTRA CUPS

6 whole cups
1 half cup
1 fourth cup
1 fifth cup
3 sixth cup

EXTRA CUPS

6 whole cups
1 half cup
1 fourth cup
1 fifth cup
3 sixth cups

TO SERVE A FULL CLASS: Prepare 2 Extra Cups gallon storage jugs. Store 12 vials in each in one, cut to size as necessary. Apply each label with clear packaging tape.

EXTRA CUPS

6 whole cups
1 half cup
1 fourth cup
1 fifth cup
1 sixth cups

FOR ONE LEARNING STATION: Prepare 1 Extra Cups gallon storage jug. Store 10 vials inside, cut to size as necessary. Apply label with clear packaging tape.

Preparation: 1 copy

film can (calibrate 6)

ruler cups (calibrate 6)

baby food jars (calibrate 4, 5, 7)

half-liter bottle (calibrate 2, 5, 7)

F / CALIBRATE
special materials

CHAPTER SIGN: Glue to a 4 x 6 inch index card, and fold in half. Stand this sign in the space where you store special materials for this chapter. Or cut this sign in half, and glue both pieces to a grocery bag that has been cut to size, or to a box. Store all listed items inside.

CUP RULERS: Fix one ruler to the side of a standard cup, clear cup, 250 ml Erlenmeyer flask, or other container of distinctive shape. Use clear packaging tape. Position so the zero marks are flush with the bottom. Trim excess ruler at the top.

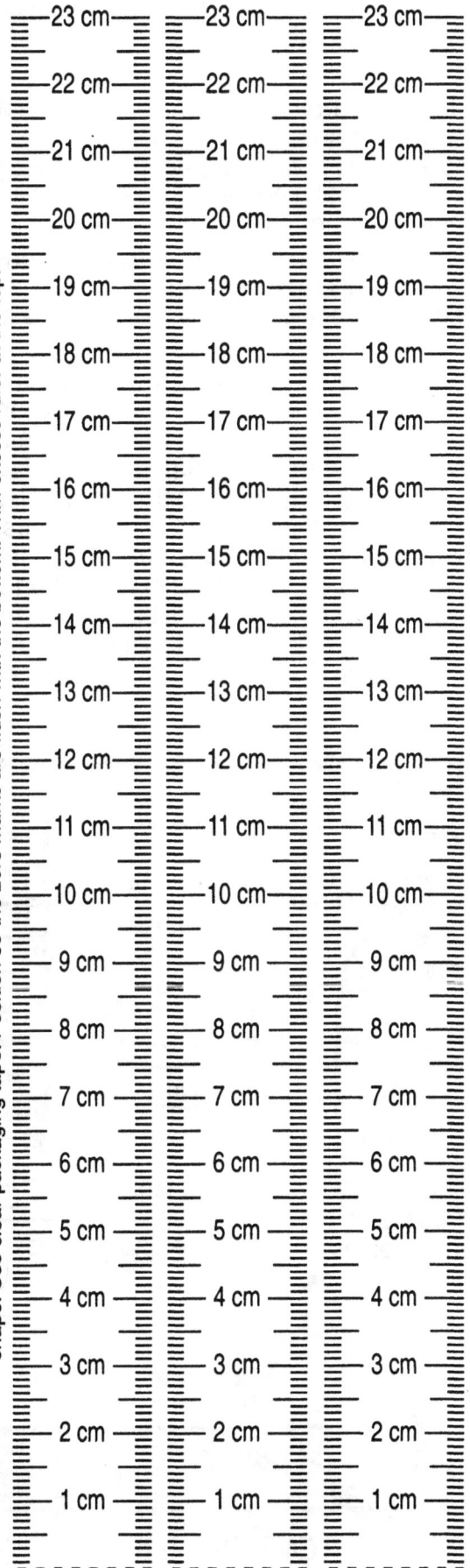

23 cm	23 cm	23 cm
22 cm	22 cm	22 cm
21 cm	21 cm	21 cm
20 cm	20 cm	20 cm
19 cm	19 cm	19 cm
18 cm	18 cm	18 cm
17 cm	17 cm	17 cm
16 cm	16 cm	16 cm
15 cm	15 cm	15 cm
14 cm	14 cm	14 cm
13 cm	13 cm	13 cm
12 cm	12 cm	12 cm
11 cm	11 cm	11 cm
10 cm	10 cm	10 cm
9 cm	9 cm	9 cm
8 cm	8 cm	8 cm
7 cm	7 cm	7 cm
6 cm	6 cm	6 cm
5 cm	5 cm	5 cm
4 cm	4 cm	4 cm
3 cm	3 cm	3 cm
2 cm	2 cm	2 cm
1 cm	1 cm	1 cm

Preparation: 1 copy

TEACHING NOTES

Purpose

To create calibrated rulers that divide a liter bottle into fractional parts.

Introduction

♦ Fill a liter bottle with lentils. Recall that "full" is a rather "squishy" term, because we can add more lentils if we compact them. We agree, by definition, that a bottle filled to the top with a loose pack is fair and full.

♦ Ask a volunteer to divide the fair-and-full liter into 2 equal parts. Will you use tubs or clear cups to hold the lentils? *Tubs. The clear cups are much too small.*

♦ Cut and fold a *Liter* Fan Ruler. Calibrate it against the height of the lentils as you funnel each equal tub back in. (Be sure to do this on the correct "page" of the ruler: the top reads "2/2 liter.")

• First tub: Mark and label "1/2 liter." Remember to keep a loose pack. "Fluff" if necessary by inverting the bottle once, then tilting it level.

• Second tub: The final mark at the top of the ruler is already calibrated, but pour in the lentils anyway. A gentle squeeze should allow any lentils remaining in the funnel to flow down to the top of the bottle.

Focus

♦ Have you prepared a *Liter* Fan Ruler?

♦ Which containers will you use to calibrate the different faces of the ruler? *Use tubs to calibrate halves and thirds. Use clear cups to calibrate fourths, sixths, and eighths.*

Checkpoint

♦ Show me your calibrated ruler:

• Is each set of calibrations marked on the correct page?

• Why is there always extra space between the top two calibrations? *Because the bottle narrows at the shoulders and neck.*

• Is each calibration fully labeled with the word "liter"? (This avoids confusion in the next job card.)

♦ Show me how to use your ruler to fill this tub with 1/3 (or 3/4, 5/6, 3/8) liter of lentils.

F/1 You need...

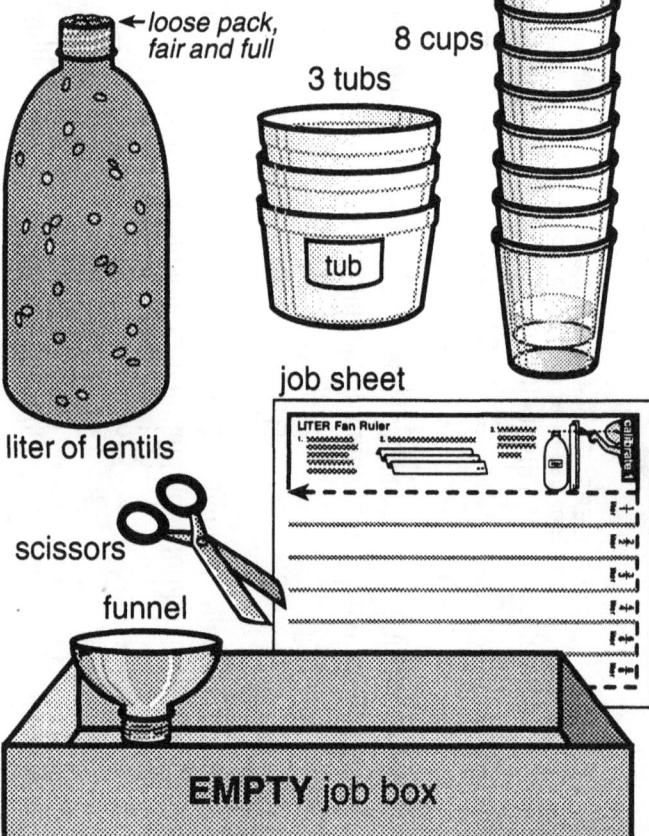

← loose pack, fair and full

8 cups

3 tubs

tub

liter of lentils

job sheet

LITER Fan Ruler

scissors

funnel

EMPTY job box

---

calibrate 1

Cut and fold a LITER Fan Ruler from the Job Sheet.

Divide a fair-and-full liter of lentils into 2 **equal** parts.

liter

tub

tub

Open the fan to the halves "page." Mark the level of a loose fill of lentils.

Label all units.

Calibrate the rest of the pages. Save your Fan Ruler.

2/2 liter

3/3 liter

liter

1/2 liter

LITER Fan Ruler

1. Cut along the dashed line on 3 sides. (Cut straight through the arrows to the edge of the paper. Do *not* cut the grey lines.)

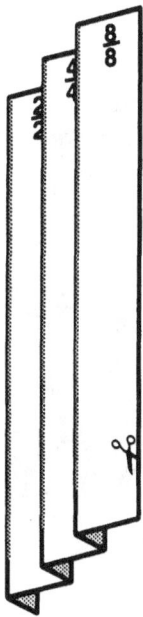

2. Fold on the grey lines like a fan.

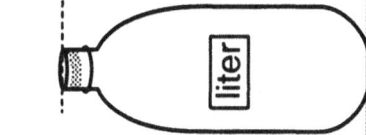

3. Stand the fan upside down beside the liter bottle. Cut it to the bottle's height.

liter

calibrate 1

$\frac{1}{1}$ liter	$\frac{2}{2}$ liter	$\frac{3}{3}$ liter	$\frac{4}{4}$ liter	$\frac{6}{6}$ liter	$\frac{8}{8}$ liter

Job Sheet: 1 copy per student

TEACHING NOTES

Purpose

To create calibrated rulers that divide a half-liter bottle into fractional parts.

Introduction

Calibrate this half-liter bottle with its own Fan Ruler, just as you did the whole liter bottle in the previous Job Card.

Focus

◆ How is this Job Card different from the last one? *We are calibrating a half-liter bottle instead of a liter bottle.*

◆ What will you do after you calibrate the half liter? *Compare measures between both bottles.* (At the lowest level, students will recognize that 2 half liters equal a whole liter. At higher levels, students will begin develop a concrete understanding of how fractions multiply together.)

Checkpoint

◆ Show me your calibrated ruler:

• Is each set of calibrations marked on the correct page?

• Is each calibration fully labeled with the word "half liter?" (This avoids confusion when comparing whole-liter and half-liter measure.)

◆ Show me the equations you have written.

1/1 half liter = 1/2 liter
1/2 half liter = 1/4 liter
1/3 half liter = 1/6 liter
2/3 half liter = 2/6 liter
1/4 half liter = 1/8 liter
3/4 half liter = 3/8 liter

◆ Can you see a pattern? *When you take half of any fraction, the denominator becomes twice as large.*

◆ What is half of a half of a half liter? *1/8 liter.*

F/2 You need...

← loose pack, filled fair and full from job box

half liter of lentils

liter bottle

4 cups

job sheet

job box with **1** liter of lentils

---

calibrate 2

Prepare a HALF LITER Fan Ruler from the Job Sheet.

Calibrate and label all its "pages" as before.

Labels are important.

Compare half liter measure to whole liter measure by pouring lentils.

1/1 *half* liter = 1/2 liter

1/2 *half* liter = ? liter

1/3 *half* liter = ? liter

2/3 *half* liter = ? liter

1/4 *half* liter = ? liter

3/4 *half* liter = ? liter

I see a pattern: A whole half is a half. A half half is a quarter. A quarter half is...

HALF LITER Fan Ruler

1. Cut along the dashed line on 3 sides. (Cut straight through the arrows to the edge of the paper. Do *not* cut the grey lines.)

2. Fold on the grey lines like a fan.

3. Stand the fan upside down beside the half liter bottle. Cut it to the bottle's height.

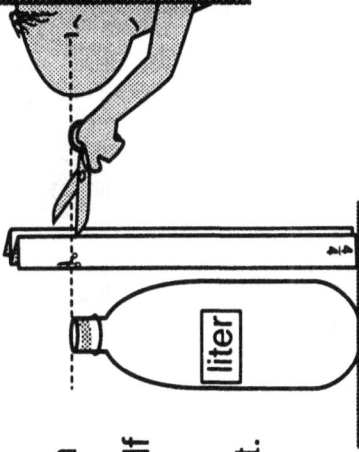

liter

$\frac{1}{1}$ half liter	$\frac{2}{2}$ half liter	$\frac{3}{3}$ half liter	$\frac{4}{4}$ half liter

113

Job Sheet: 1 copy per student

TEACHING NOTES

Purpose

To calibrate a liter bottle in cups and half cups. To design an attractive label.

Introduction

You work for a company that sells Fizzy Soda. Business has been slumping lately, so the boss asks you to redesign the liter bottle that Fizzy Soda is sold in, to make it more useful and attractive.

She wants you to calibrate it in cups and half cups, so customers will know how much soda remains in bottles they have already opened. She also wants you to draw a bold new label that consumers will notice on the grocery shelf.

Focus

◆ How big will you make your label? Does it matter where you stick it on the bottle? (If the label is not placed at the correct height, the calibrations will make no sense. It may be convenient to line up the bottom of the label with its lowest horizontal ridge or seam. On liter bottles we have used, this seam appears just below the 1 cup level.)

◆ Will you use masking tape as part of your design?

Checkpoint

Possible designs:

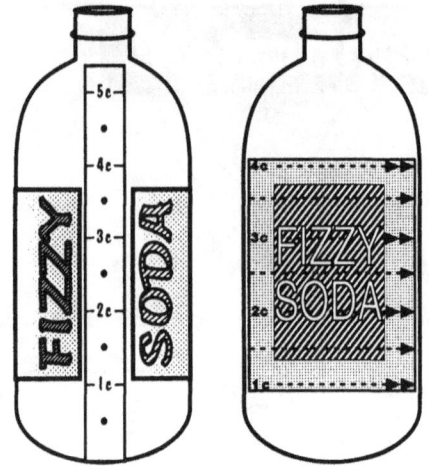

◆ Show me how to measure __ cups with this bottle.

◆ Draw this bottle full size. Paste your label and calibrations to your drawing in the correct position.

F/3 You need...

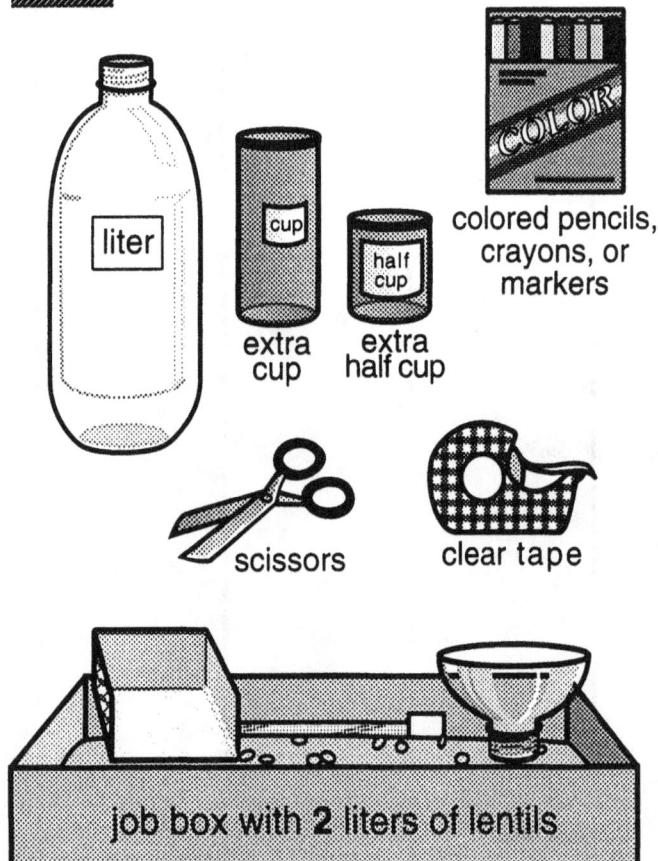

liter

extra cup

extra half cup

colored pencils, crayons, or markers

scissors

clear tape

job box with **2 liters** of lentils

---

calibrate 3

Calibrate our liter bottle in cups and half cups...

...and design a better label.

FIZZY SODA

TEACHING NOTES

Purpose

To calibrate and graph a liter bottle in BFJ increments of different size.

Introduction

◆ Run a long strip of masking tape from the "neck" of the bottle all the way down under its "foot." Keep the tape on the roll until you cut it off a little below the foot.

◆ Draw a bold baseline across the tape where it meets the table. Label it "0."

◆ As a quick demonstration, calibrate the masking tape in whole cups, numbering them 1, 2, 3, and 4. Remind students to maintain a loose pack. If the lentils settle, invert once and tilt to level.

◆ Peel the calibrated tape from the bottle. Stick it to one of the 3 grey strips on the Ruler Paper so *baseline zeros match.*

◆ Notice how each numbered line from 0 to 4 reaches to a different height in centimeters. Plot these ordered pairs on the Graph Paper, beginning with (0,0) which is already marked. Circle the points and draw a graph line.

◆ This graph line shows how the height of lentils in a liter bottle changes with added cups. How might it look different when you calibrate it instead in BFJ's? *The smaller containers will cause the lentils to rise less per addition, resulting in shallower graph lines.*

Focus

◆ How will you prepare the liter bottle for calibrating? *Stick masking tape down the side; draw a bold baseline at the very bottom, and number it 0.*

◆ You will be graphing three sizes of BFJ's. How many sheets of Ruler Paper and Graph Paper will you need? *Just 1 of each sheet. I'll stick all 3 calibrated tape strips to the same Ruler Paper, and draw and label all three graph lines on the same Graph Paper.*

Checkpoint

◆ What shape are your graph lines? *Straight, with some scattering due to error.* (Very careful students may notice an upswing as the bottle narrows at the top. This will only be noticed if loose fill is maintained up into the shoulders and neck of the bottle, so accurate points can be plotted at the top of the graph, even beyond.)

◆ Compare/Contrast: *All 3 BFJ's graph as nearly straight lines, with the smallest jar producing more closely spaced calibrations, and therefore the shallowest graph line.*

F/4 You need...

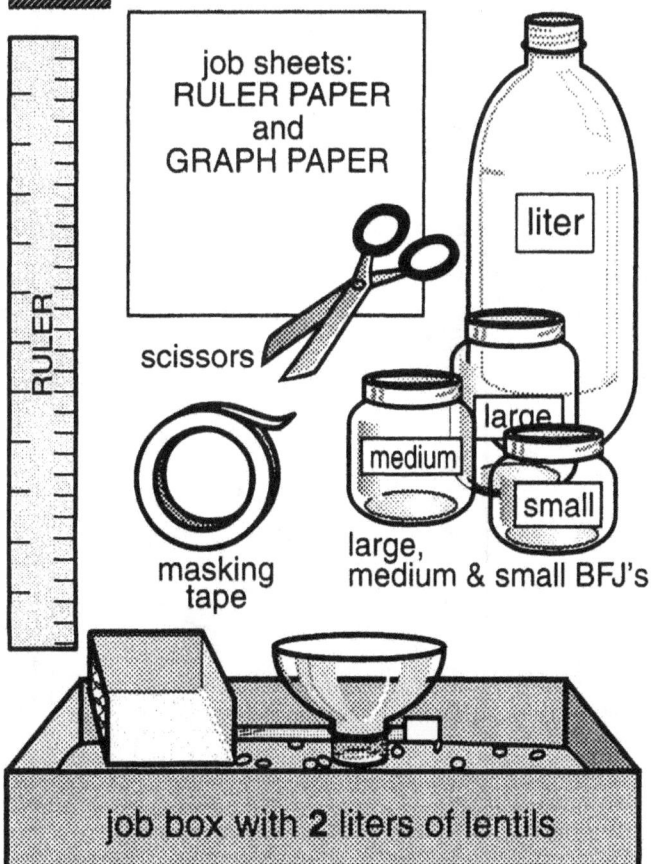

job sheets:
RULER PAPER
and
GRAPH PAPER

liter

scissors

large

medium

small

masking tape

large, medium & small BFJ's

RULER

job box with 2 liters of lentils

---

calibrate 4

1. Calibrate a liter bottle in large BFJ's. Begin with zero at the bottom.

2. Peel off the tape. Stick it to RULER PAPER so baseline zeros match.

3. Plot, circle and label each ordered pair on GRAPH PAPER.

large

Number.

liter

3
2
1
0

BASELINE

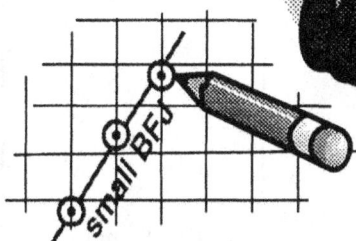

small BFJ

medium   small

Repeat steps **1-3** for medium and small BFJ's.

Ruler Paper
Centimeters

23
22
21
20
19
18
17
16
15
14
13
12
11
10
9
8
7
6
5
4
3
2
1

BASELINE — 0

0   0   0

Compare. How are calibratioins and graph lines similar? Different?

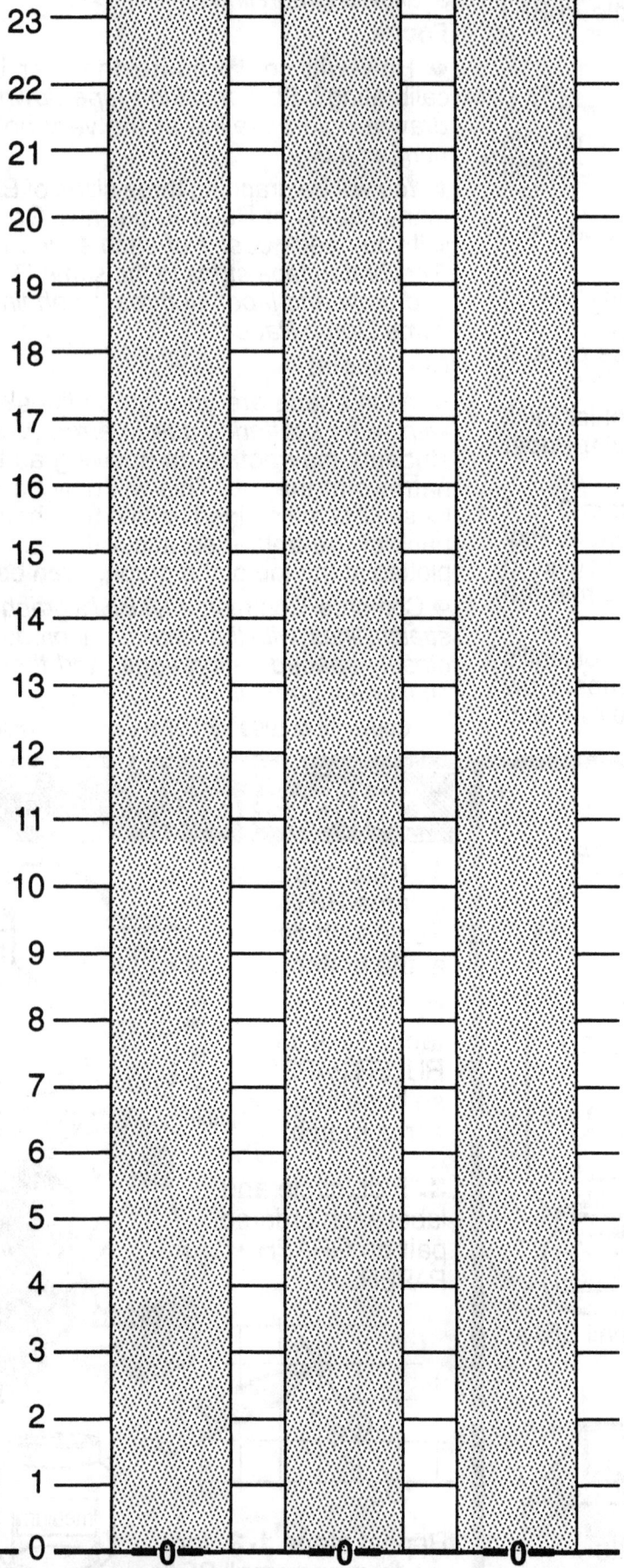

REMEMBER TO DO THIS PART.

Job Sheet: 1 copy per student

Graph Paper

calibrate 4, 5, 6

Height of Lentils (centimeters)

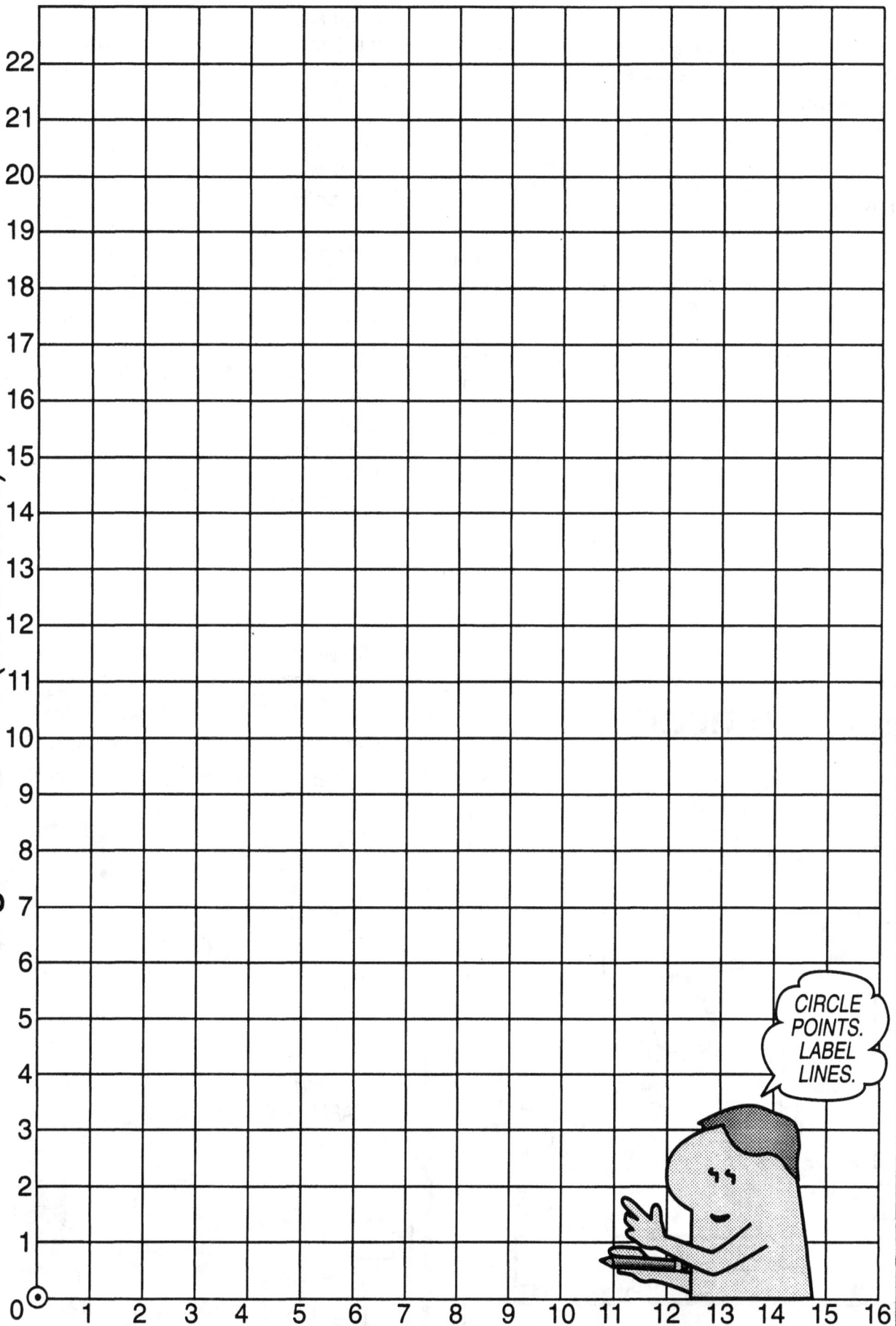

22
21
20
19
18
17
16
15
14
13
12
11
10
9
8
7
6
5
4
3
2
1
0

1 2 3 4 5 6 7 8 9 10 11 12 13 14 15 16

Number of Containers (fair and full)

CIRCLE POINTS. LABEL LINES.

Job Sheet: several copies per student

TEACHING NOTES

Purpose

To calibrate and graph a half-liter bottle in BFJ's and fraction cups of different size.

Introduction

◆ Last time you calibrated and graphed a whole liter bottle (hold one up). For this job card you will calibrate and graph a half liter bottle (hold one up). Notice the difference in diameters.

◆ Predict how graph lines for BFJ's might look different when calibrated up this narrow half-liter bottle. *The half liter should fill faster, causing the graph lines to rise more steeply.*

◆ After doing the 3 BFJ's, you will calibrate and graph these 3 smaller fraction cups on the same graph paper (hold them up). What might a graph with 6 lines look like? *They all start at (0,0). The lines will rise less steeply as the containers get smaller. Perhaps they will radiate like spokes in a wheel.*

Focus

◆ What 6 containers will you be graphing on the same Graph Paper? *From large to small: large BFJ, medium BFJ, small BFJ, 1/4 cup, 1/5 cup and 1/6 cup.*

◆ How much Ruler Paper will you need? *Two sheets: 1 for the 3 BFJ's, and another for the 3 fraction cups.*

Checkpoint

◆ How do your graph lines appear? *Straight; radiating from (0,0) like spokes in a wheel, with some scattering due to error.* (Again, some upswing may be visible if loose fill is maintained up into the shoulders and neck of the bottle. This is especially true for the smaller fraction vials, if points are plotted accurately at the top of the graph.)

◆ Compare / Contrast: *All 6 lines are nearly straight, with the smallest container producing more closely spaced calibrations, and therefore the shallowest graph line.*

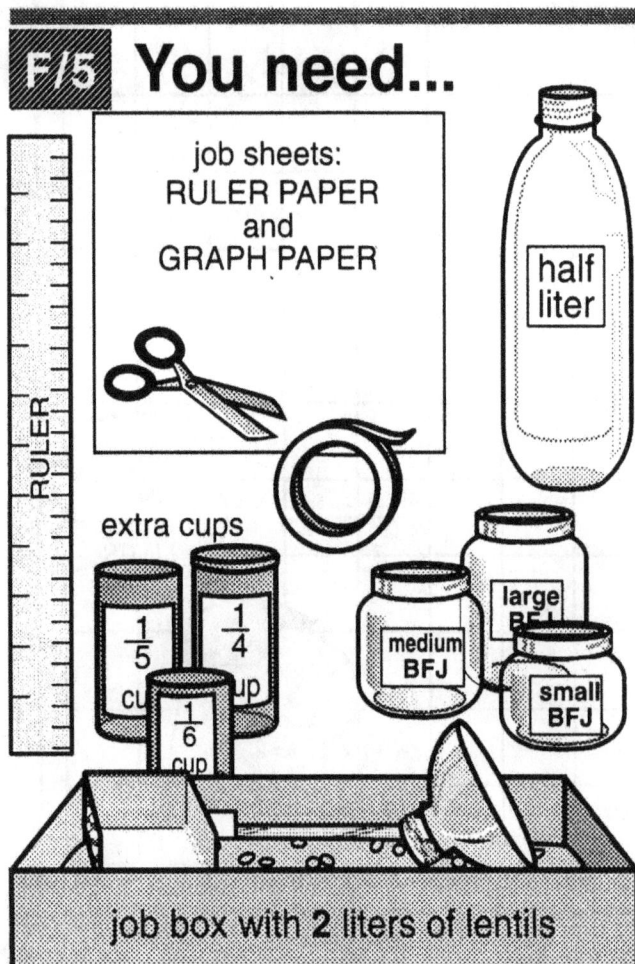

F/5 You need...

job sheets:
RULER PAPER
and
GRAPH PAPER

RULER

half liter

extra cups

$\frac{1}{5}$ cup $\frac{1}{4}$ cup $\frac{1}{6}$ cup

medium BFJ   large BFJ   small BFJ

job box with **2 liters of lentils**

calibrate 5   👤 or 👤👤

Calibrate a HALF liter bottle in BFJ's as you did the liter bottle.

large BFJ   medium BFJ   small BFJ

Use new RULER PAPER and GRAPH PAPER.

1

0

$\frac{1}{4}$ cup $\frac{1}{5}$ cup $\frac{1}{6}$ cup

Repeat, with fraction cups. Use new Ruler Paper but add to the *same* Graph.

Purpose

To generate data tables, plot ordered pairs, and graph results.

Introduction

◆ Draw a large scale divided into tenths. Keep about a chalk width between divisions. Point to these with a pencil, reciting measurements aloud: 0.0 units, 0.1 units, 0.5 units, 0.9 units, 1.0 units, 1.7 units, etc.

◆ Fill a ruled container with any portion of lentils. Invite students, one at a time, to read the scale and write on scratch paper how high they think the lentils reach inside. List these measurements. Discuss the nature of measuring certainty:

• If you were asked to measure the height of a crowd, would you choose the tallest person, or try to find some average height among all those heads? Does it seem OK that others might get a slightly different answer than you?

• Which numbers on this list seem just a little different, and which seem to be wrong? Cross out the mistakes.

• Draw a wavy side view of a lentil surface on your board. An average height might be about here: 〰️

Focus

◆ Which container will you try first?

◆ What will you do? *Pour. Measure. Record. Plot. Connect. Label. Interpret.*

Checkpoint

◆ Show me your data tables and graphs. (Are ordered pairs accurately plotted and circled? Are graph lines smooth, seeking an average among the scattered points? Is each graph line labeled with the cup that generated it?)

◆ Interpretations: (container/graph line)

(narrow/steep)　　　(widening/curving: less
(wide/shallow)　　　　　　　　　　steep)
(straight/straight)　(narrowing/curving: more
　　　　　　　　　　　　　　　　　　steep)

More

◆ Invite students to bring their own clean and dry novelty bottles for graphing. They must be transparent, and if glass, not too thickly distorted.

◆ Can experimental accuracy be increased by using water instead of lentils? (Limit any water experiments to an area absolutely separate from the lentils. There is wonderful science in sprouting lentils, but this is subject matter for another book.)

F/6 You need...

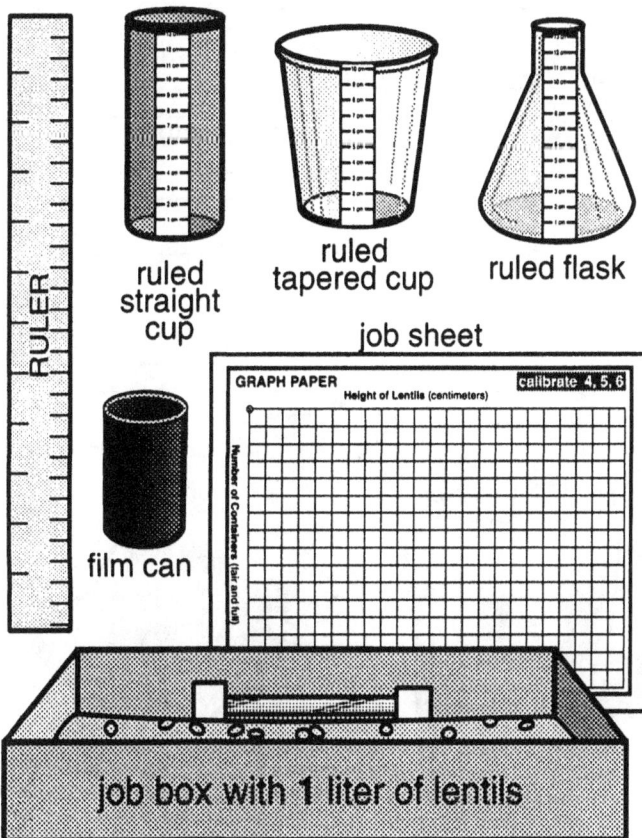

ruled straight cup

ruled tapered cup

ruled flask

job sheet

GRAPH PAPER
Height of Lentils (centimeters)　　calibrate 4, 5, 6
Number of Containers (fair and full)

film can

job box with **1 liter of lentils**

---

calibrate 6　👤 or 👤👤

Add film cans of lentils to each ruled cup.

FAIR AND FULL

Record each level in a table:

NUMBER OF CANS	HEIGHT (cm)
0	0
1	
2	
3	

Graph your data. How does the shape of the container affect the graph line?

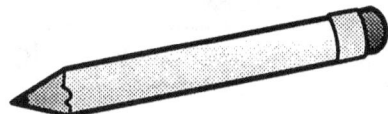

TEACHING NOTES

Purpose

To provide students with an approved way to pursue their own ideas about calibrating and graphing volumes.

Introduction

Display this card whenever you want to do this suggested activity or experiments that you design yourself.

Focus

◆ What do you want to study about calibrating and graphing?

◆ Are these the materials you need?

Checkpoint

Report in words and pictures:
- What you did.
- What you learned.
- Questions you may still have.

Special Note

Recognize excellence: Record some of your students' most creative ideas on the opposite page.

F/7 **You need...**

clear cup

masking tape

DIVIDE 10 containers

half liter

liter

large BFJ

medium BFJ

small BFJ

job box with **2 liters of lentils**

☞ Ask your teacher for other items. Tell why you need them.

---

calibrate 7 👤 or 👤👤

On your own.

○ My idea:

Calibrate a measuring cup.

○

whole cups

half cups

third cups

G / ESTIMATE

In this chapter: Spread lentils on a Counting Grid to form an array. Multiply to know their exact number. This is a fast and easy way to count lentils in a bottle cap or establish the height of 100 lentils in a test tube. Use these relationships to estimate the number of lentils in much larger volumes. Will a liter bottle hold 2^{14} lentils if you pack them in very tightly? How many times can you divide these lentils in half before you end up with only one? Count the number of lentils that fill a cubic inch. Can you estimate how may lentils fill a Job Box? Will it hold a million?

Basic Materials: Quantities define maximums needed to support any one Job Card in this chapter. Store *high quantity basics* (Job Boxes, liters of lentils, bottle lids, scoops, funnels) on and under a table or counter. Store *low quantity basics* near the "basics" sign (see page 45) or in a "basics" box. Consult our Glossary on pages 6-9 for a full description of these items. See the next page for additional special materials used in this chapter.

- ☐ **1 job box**
- ☐ **up to 3 liters of lentils**
- ☐ **bottle cap spoon**
- ☐ **1 tub**
- ☐ **masking tape**
- ☐ **extra cup**
- ☐ **1 craft stick**
- ☐ **stick rulers**
- ☐ **clear tape**

☞ ***Please observe our copyright restrictions on page 2.***

Store these chapter-specific items together in a designated place. They require about 1 square foot of dedicated space. General classroom materials (like scissors and tape) are also listed below when used, while others (like pencil and paper) are always assumed.

spreader (estimate 1, 2, 3, 4, 5, 9)

Cut a square of **index card** about the size of a postage stamp. Stick it to the end of a **craft stick** with glue or rolled masking tape so about half of it projects beyond the stick.

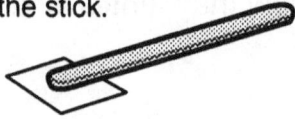

counting grid (estimate 1, 2, 3, 4, 5, 9)

Cut this from the same $1/4$ **inch hardware cloth** used to make sorting screens in Search. Use **wire cutters** to cut a full 3 x 3 inch piece (12 little squares along each side), *plus* an extra fringe of half squares all around the perimeter. Center this square grid of wire on **any smooth, rigid material** big enough to hold the whole screen. This might be a square of ceramic tile, or glass, or foam board, or Masonite, or even a large, inflexible plastic or metal lid. Press the wire into a very shallow bowl to assure good, firm contact with the support. Tape it firmly to the support using $3/4$ inch plastic **electrical tape**, covering the wire fringe plus one row of squares all around. This leaves a 10 x 10 grid of wire squares uncovered in the middle.

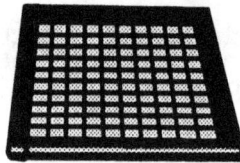

Caution students to treat these counting grids with care and respect. If dropped, they could break. *Never* pry up on the middle of the screen to dislodge a stuck lentil. (Rather, give it a sharp slap from behind.) Once separation occurs between screen and backing, the grid must be rebuilt.

test tube (estimate 2, 3)

Use a small to medium size with an inside diameter of about 1/2 inch. Narrow is best, as long as the tube can hold at least 100 loosely-packed lentils.

SPECIAL NOTE: This chapter borrows 3 specific items dedicated for use in other chapters. If you wish to avoid borrowing across chapters, duplicate extra sets of these loaned materials for exclusive use here.

set of 10 fraction cups (estimate 4, 5, 7, 9)

Borrow this from Divide.

set of 6 measuring cups (estimate 4)

Borrow this from Measure.

halving pattern (estimate 6)

Photocopy the line master on page 130 and trim along the dotted lines. This gives you a pattern sized to a Job Box with a bottom area of 19 x 14.5 inches (49 x 37 cm). Lay this trimmed pattern in a bottom corner or your particular box to confirm that it covers a fourth of the total "floor." Even an approximate match is probably good enough. Otherwise, resize as necessary.

set of 3 rocks and 3 cards (estimate 7)

Borrow this from Divide.

cardboard dam (estimate 7)

Cut a rectangle of **corrugated cardboard** equal in width and height to the end of your job box. Ours measured 3 x 14.5 inches (8 x 37 cm).

inch grid paper (estimate 8)

Photocopy the Cross Patterns on page 133. Divide into six pieces per sheet.

calculator (estimate 8, 9)

Optional.

blunt scissors (estimate 8, 9)

short quart carton (estimate 9)

Cut to size so it stands 4 cm high.

short quart carton (estimate 9)

inch grid paper (estimate 8)

cardboard dam (estimate 7)

halving pattern (estimate 6)

test tube (estimate 2, 3)

counting grid (estimate 1, 2, 3, 4, 5, 9)

spreader (estimate 1, 2, 3, 4, 5, 9)

G / ESTIMATE
special materials

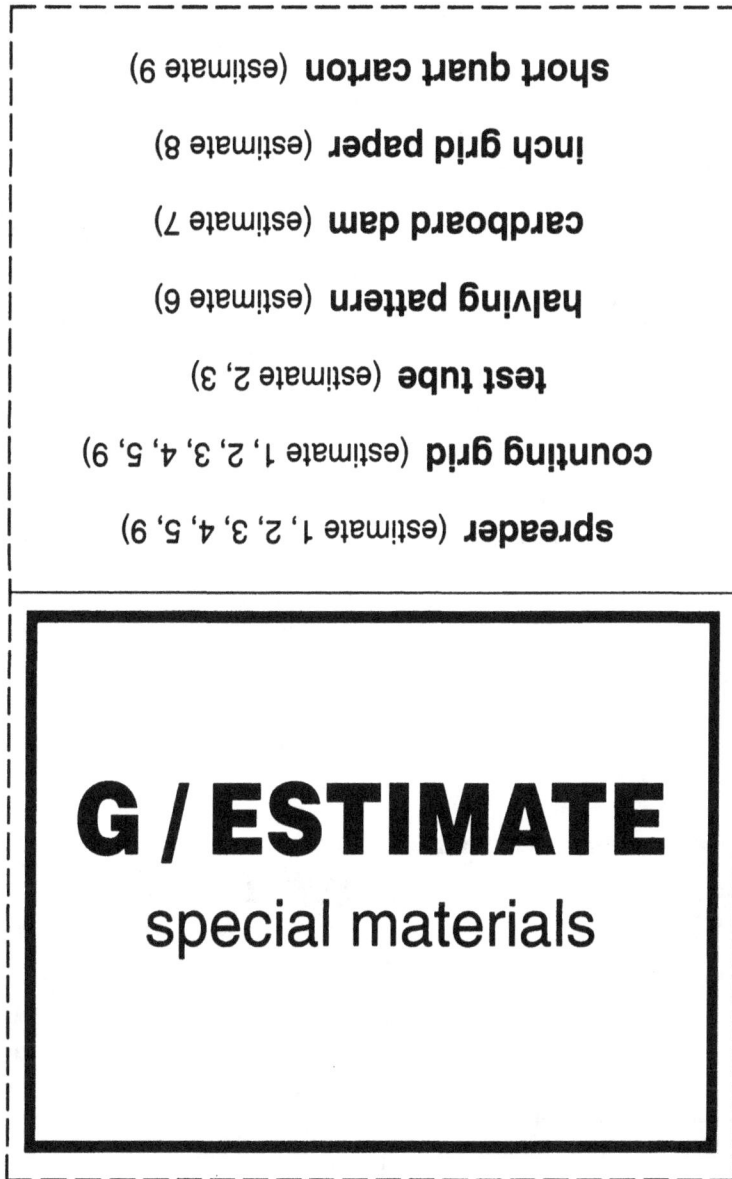

CHAPTER SIGN: Glue to a 4 x 6 inch index card, and fold in half. Stand this sign in the space where you store special materials for this chapter. Or cut this sign in half, and glue both pieces to a grocery bag that has been cut to size, or to a box. Store all listed items inside.

Preparation: 1 copy

TEACHING NOTES

Purpose

To count the number of lentils in a bottle cap by forming an array on a wire grid. To find the mode, median or mean in a group of numbers.

Introduction

◗ How to use the Counting Grid and Spreader:

• Hold the Spreader almost flat. Turn it so the full craft stick remains visible above the card.

• Spread the lentils, with the edge of the card, across the grid like butter on bread.

• Move lentils with a corner of the card. Nudge them into adjacent squares, or scoop them up to carry across the grid.

• Count by multiplying length and width of the array. Then add the few lentils that may be left over, or subtract the few empty spaces.

• Never pry up on the screen to remove a stuck lentil. Once the screen has been raised, lentils will sneak underneath again and again. Rather, give the grid a sharp rap from behind, or ask me (the teacher), to remove it.

◗ A fair and full bottle cap usually spreads over the counting grid in 1 of 4 arrays; write these multiplication facts on the board for review:

$$6 \times 8 = 48 \qquad 7 \times 7 = 49$$
$$6 \times 9 = 54 \qquad 7 \times 8 = 56$$

◗ How to decide the *accepted value*: Throw out any totals that seem to fall outside the normal range, and count again. Select among *at least* 5 trial numbers in any of the following ways:

• Take the **mode**: Find the number that appears most often in the group.

• Take the **median**: Arrange the numbers in a row from small to large. Pick the middle number in this row.

• Take the **mean** (average): Find the sum of all the trials and divide it by the number of trials.

Focus

◆ Practice taking fair and full bottle caps so you can capture almost the same number each time.

◆ How will you count? *Make an array on the grid.* (Some may simply spread the lentils over the grid, then count them row by row. This requires less effort to arrange, and less thought, but more time to count. Overall, it is a longer, less efficient process.)

◆ How will you decide on the accepted value? *Find the mode, median, or mean over 5 valid trials.*

Checkpoint

Show me your math. Tell me what you did. (A fair and full bottle cap equals 51 ± 3 lentils.)

G/1 You need...

spreader

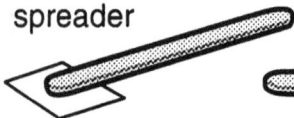

bottle-cap spoon

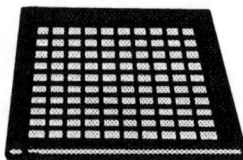

counting grid

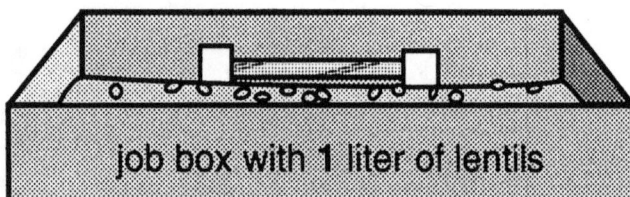

job box with 1 liter of lentils

---

estimate 1 🧍

Overfill a Bottle-Cap Spoon. Gently shake off excess lentils so it is fair and full.

Spread this on the grid, in columns and rows of single lentils. Find the total number.

Repeat at least 5 times: How did you choose your accepted value?

Trial 1:	
Trial 2:	
Trial 3:	
Trial 4:	
Trial 5:	
Accepted value:	

TEACHING NOTES

Purpose

To estimate number by sampling a constant volume. To experience experimental uncertainty and work to minimize it.

Introduction

▶ Cover the grid with lentils and shake it gently and continuously from side to side. Keep shaking until only a single layer of lentils remains and empty squares begin to appear in the 10 x 10 array. This is the most efficient stopping point. It is easier to fill these few gaps with new lentils than to stop shaking sooner and clear away many extra lentils.

▶ How to calibrate the 100-lentil level:

• Complete the 10 x 10 array. Dump it into the wide tub, then carefully pour the lentils into the narrow test tube, using your fingers as a funnel.

• Apply a short piece of vertical masking tape to the side of the test tube at the approximate height of the lentils inside.

• Make 2 tick marks in pencil along an edge of the tape: The *higher mark* corresponds to a *very loose pack* of 100 lentils allowed to slide down the side of the tube and stack up as high as possible; the *lower mark* corresponds to a *very tight pack* of 100 lentils, shaken together so the lentils rest as low as possible.

• Check these higher and lower tick marks several times by loosening and then resettling the contents. It's OK to add new pencil marks as you change your mind. When you are certain, mark the "official" high and low levels in ink.

Focus

◆ How will you know you have exactly 100 lentils in the test tube? *Make a 10 x 10 array on the grid.*

◆ Why make 2 calibrations? *A higher one marks very loose fill; a lower one marks very tight fill.*

◆ What sort of marks will you make on the tape? *Little tick marks in pencil along the edge.*

◆ Why use both pencil and ink? *Use pencil for experimental marks, then ink for final results.*

◆ How will you test your calibrated tube for accuracy? *Sample lentils to the calibration lines, then count them on the grid.*

Checkpoint

◆ How accurately did you estimate with the calibrated test tube? (100 ± 3 accuracy is possible.)

◆ Advantages / Disadvantages: Estimating is easier and faster, but only approximate. Counting is exact, but tedious and time consuming.

G/2 You need...

masking tape

tub

test tube

counting grid

pencil and pen

spreader

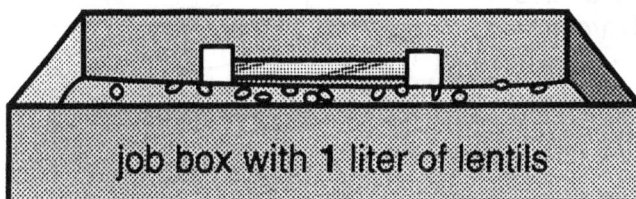

job box with 1 liter of lentils

---

estimate 2

Count exactly 100 lentils with the Counting Grid. Mark their level in a test tube.

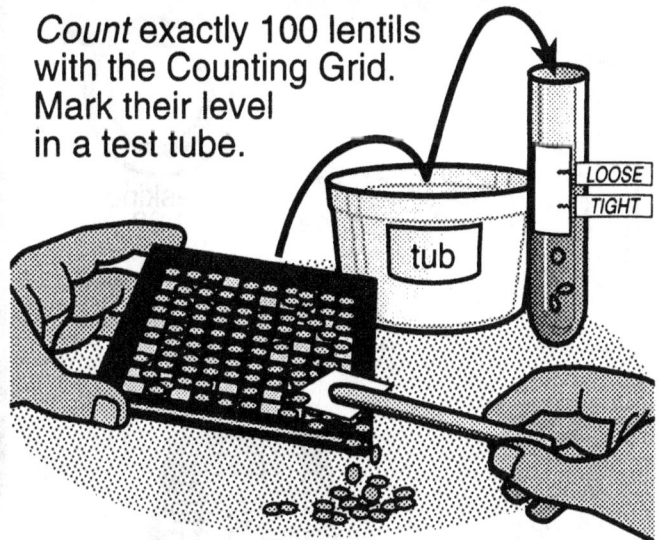

LOOSE
TIGHT
tub

Estimate 100 lentils in the empty test tube. Check how close you got on the counting grid. Repeat at least 3 times.

Trial 1:	
Trial 2:	
Trial 3:	

List advantages and disadvantages for estimating. For counting.

TEACHING NOTES

Purpose

To calibrate 1/6 cup in 100-lentil increments. To confirm the accuracy of these calibrations by different methods of counting and estimating.

Introduction

You will be calibrating this 1/6 cup in increments of 100 lentils. What different ways might you do this?

- Arrange a 10 x 10 array of lentils on the counting grid, and pour them in 100 at a time.
- Add fair-and-full portions to a Bottle-Cap Spoon. Pour them in 2 at a time, adding or subtracting a few lentils, as required, to estimate 100.
- Fill the cup with portions from a test tube

calibrated to the 100 lentil level.

Focus

◆ What is your mission? *To calibrate the 1/6 cup in 100-lentil increments.*

◆ How will you accomplish it? *In one of 3 ways: adding lentils counted on the grid; measuring in the bottle cap; measuring in the test tube.*

Checkpoint

◆ May I see your report?

◆ How did you check the accuracy of this calibrated tape? *I tested it using a different method than I used to calibrate it.*

More

◆ Add lentils by fair-and-full bottle-cap spoons to your calibrated test tube. Does one estimate confirm the other?

◆ Partly fill a counting grid with lentils.
- Try to estimate the percent of filled and empty spaces just by looking.
- How close was your estimate? Can you improve with practice?

G/3 You need...

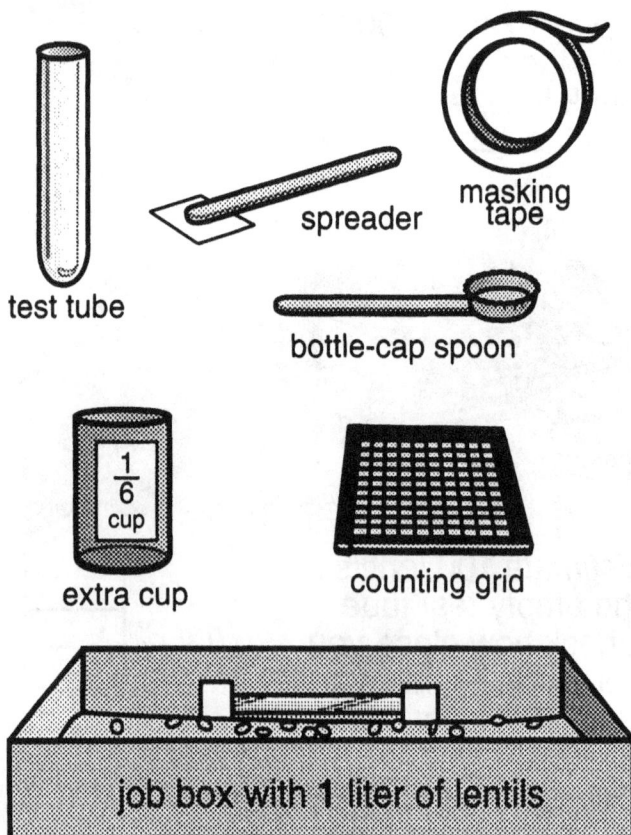

test tube

spreader

masking tape

bottle-cap spoon

1/6 cup

extra cup

counting grid

job box with 1 liter of lentils

estimate 3

Stick masking tape to the side of a 1/6 cup. Mark the baseline.

Calibrate in 100 lentil increments (loose pack).

Confirm the accuracy of your work by other methods.

COUNTING GRID?
BOTTLE CAP ESTIMATE?
TEST TUBE ESTIMATE?

—30
—200
—100

BASELINE

Write a report. Include a full-sized side view of the 1/6 cup, with your calibrated tape stuck on it.

500
400
300
200
100

TEACHING NOTES

Purpose

To calculate the number of lentils in large and small volumes, based on an estimate of 500 lentils in 1/6 cup. To confirm the accuracy of these calculations in smaller volumes by actual count.

Introduction

In this job card we assume that 500 lentils will fill 1/6 cup, then calculate how many lentils all the other cups hold based on this estimate.

• What if this estimate is too low? *All other estimates based on it will also be too low.*

• What if this estimate is too high? *All other estimates based on it will also be too high.*

• What if this estimate is really close? *Estimates based on it will be reasonably close.* (Whatever the uncertainity may be, it multiplies with increasing volume, and divides with decreasing volume.)

Focus

◆ How will you organize your work? *In a table.* (Students who get lost in a maze of calculations might try a more direct approach: Calculate and tape the totals on one cup at a time. Construct a table as a final step, and remove the tape labels.)

◆ How will you evaluate the accuracy of your calculated estimates? *Count the lentils in small fair-and-full volumes, with the Counting Grid.*

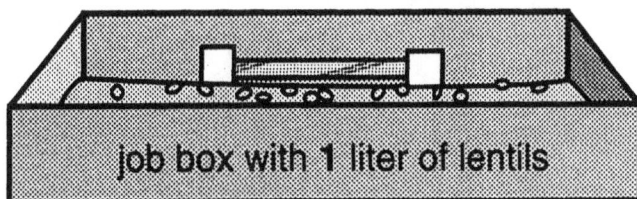

Checkpoint

◆ May I see your calculations and table of estimates?

$1/6$ C = 500	1 pt = 6,000	$1/8$ C = 375
$1/3$ C = 1,000	1 tub = 9,000	$1/10$ C = 300
$1/12$ C = 250	1 qt = 12,000	$1/5$ C = 600
1 C = 3,000	$1/2$ C = 1,500	$1/20$ C = 150
	$1/4$ C = 750	

◆ Do you think these estimates are accurate? (Students should form opinions based on actual counts. We have found these to vary by ± 4% from calculated estimates.)

More

◆ Calculate your percent of error between estimated and exact counts:

$$\% \text{ error} = \frac{\text{difference}}{\text{actual count}} \times 100$$

◆ Construct a more accurate table of estimates based on numbers that are less round.

◆ Estimate the volume occupied by…
ten thousand lentils (10,000): 3.3 cups;
a hundred thousand lentils (100,000): 33.3 cups;
a million lentils (1,000,000): 333.3 cups.

G/4 You need...

DIVIDE 10 containers

MEASURE 5 containers

spreader

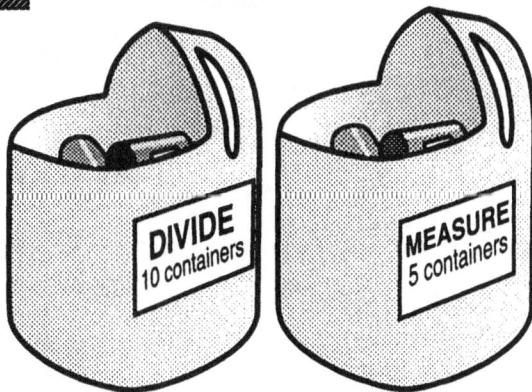

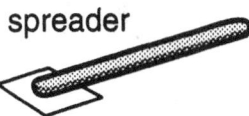

counting grid

job box with **1** liter of lentils

---

estimate 4 👤 or 👤👤

Multiply to estimate the lentils in each container:

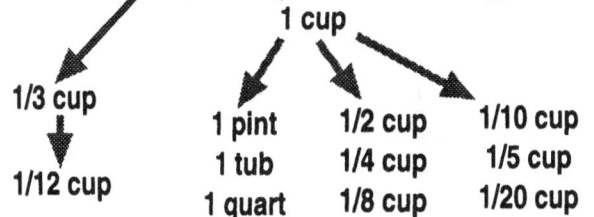

Estimate this container to hold **500 lentils**

1/6 cup

1 cup

1/3 cup

1/12 cup

1 pint
1 tub
1 quart

1/2 cup
1/4 cup
1/8 cup

1/10 cup
1/5 cup
1/20 cup

Show your math in an organized way.

Save your estimates in a reference table.

Count lentils in some smaller cups to evaluate the accuracy of your round-number estimates.

TEACHING NOTES

Purpose

To explore number patterns. To experience exponential doubling.

Introduction

♦ Write this pattern of multiplication (right side only) on your board. After the second line, let students tell you how to complete each line.

START ON THIS SIDE:

$2^1 = 2 = 2$
$2^2 = 2 \times 2 = 4$
$2^3 = 2 \times 2 \times 2 = 8$
$2^4 = 2 \times 2 \times 2 \times 2 = 16$
$2^5 = 2 \times 2 \times 2 \times 2 \times 2 = 32$

After establishing the pattern on the right up to 32, introduce the pattern to the left. The small raised numbers are called *exponents*. They tell you how many times to multiply the base number.

♦ Erase the board. Work a few more problems to reinforce the the use of exponents:

$2^4 = ?$ $2^3 = ?$ $2^5 = ?$ $3^2 = ?$ $3^3 = ?$

Focus

◆ How far will you extend your table? *Up to 2^{14}.*

◆ How many lentils will you try to fit into the liter bottle? *2^{14}.*

Checkpoint

◆ May I see your table of 2's?

$2^0 = 1$	$2^4 = 16$	$2^8 = 256$	$2^{12} = 4{,}096$
$2^1 = 2$	$2^5 = 32$	$2^9 = 512$	$2^{13} = 8{,}192$
$2^2 = 4$	$2^6 = 64$	$2^{10} = 1{,}024$	$2^{14} = 16{,}384$
$2^3 = 8$	$2^7 = 128$	$2^{11} = 2{,}048$	

(Why does a number with an exponent of zero equal 1? Because it works! It fits the pattern in the table. Or ask how many twos multiply together to get 1? Answer: No twos! 0! More advanced students might follow $2^3 / 2^3 = 2^{3-3} = 2^0$.)

◆ Did 2^{14} lentils (approximately) fit into a liter bottle? Show me your math. *Yes, 16,384 will fit if you really pack them down* (tickle, shake, poke).

5 cups = 5 × 3,000 = 15,000 lentils
1/3 cup = 1,000
1/10 cup = 300
counting grid = 84
 16,384 lentils

More

◆ Is 2^{20} more than 1 million (1,000,000)? *Yes, about 5% more.* (Simplify by grouping multiples: $2 \times 2 \times 2 \times 2$ is the same as 4×4, is the same as 16.)

◆ If 2^{14} lentils fill 1 liter bottle, how many liter bottles are filled by 2^{20} lentils? *64 liter bottles.*

G/5 You need...

...the table of estimates you made for job card 4.

spreader

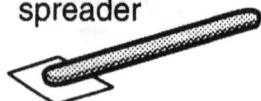

counting grid

DIVIDE 10 containers

EMPTY job box

---

estimate 5 🕴 or 🕴🕴

Complete this table of exponential 2's. Keep multiplying as far as 2^{14}.

Will 2^{14} lentils fit into a liter bottle? Explain how you figured this out.

2^0	1
2^1	2
2^2	4
2^3	8
2^4	16
2^5	
↓	
2^{14}	

2^{14} lentils ?

liter

TEACHING NOTES

Purpose

To explore number patterns. To experience exponential halving.

Introduction

▶ Begin with an empty job box and 2 liters of lentils. Pour part of one bottle into the other until it is very tightly packed. The total number of lentils inside is now close to what special number? *2^{14}, or 16,384.*

▶ Fill the empty Job Box with 2^{14} lentils. Pour them evenly across the bottom of the box, including all 4 corners. Give the box a sharp shake to create a uniform distribution of lentils.

▶ Place the Halving Pattern in a corner of the Job Box so it covers a quarter-section of lentils.

• Using this paper as a visual guide, push *half* the lentil "floor" away to one side of the box with the craft stick. Estimate the number of lentils that remain in the undisturbed half. *16,382/2 = 8,192.*

• Divide again. Push *half* of the remaining floor (the uncovered part) away. How many lentils remain under the paper? *8,192/2 = 4,096.*

• How many times can we divide this floor before we are left with just 1 lentil? Encourage debate. Let the mystery hang.

Focus

◆ How many lentils will you pour into your Job Box?

◆ How will you arrange and divide them?

Checkpoint

◆ How many times can you divide a liter of lentils in half? *14 times.*

◆ Let me see your table. Why does it start with zero? *Zero means no divisions. You are starting with the whole undivided floor.*

1/1 = 16,384	*1/32 = 512*	*1/1,024 = 16*
1/2 = 8,192	*1/64 = 256*	*1/2,048 = 2*
1/4 = 4,096	*1/128 = 128*	*1/4,096 = 4*
1/8 = 2,048	*1/256 = 64*	*1/8,192 = 2*
1/16 = 1,024	*1/512 = 32*	*1/16,384 = 1*

(Notice that there are 15 table entries, but only 14 divisions.)

More

Imagine that you don't stop at 1 lentil, but keep on dividing: 1/2 lentil, 1/4 lentil, 1/8 lentil…. Would you ever reach a point where there is nothing left to divide? What is the smallest part of a lentil? Is there a smallest part?

G/6 You need...

Pack as much as possible into one bottle.

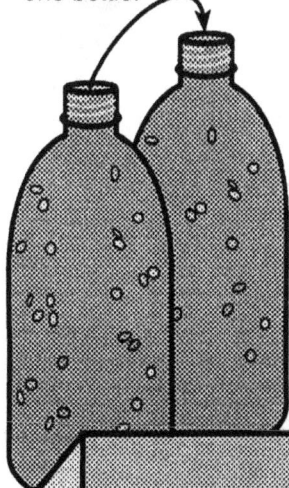

halving pattern

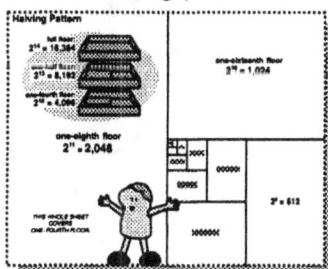

2 liters of lentils

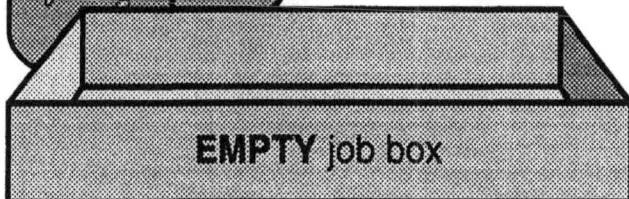

craft stick *(for halving lentils)*

EMPTY job box

---

estimate 6 🧍 or 🧍🧍

Spread 16,384 lentils evenly across your job box. Shake them to make a level "floor."

Measure with the Halving Pattern.

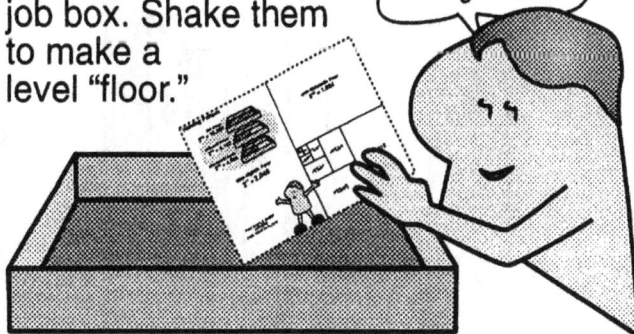

Starting with zero divisions (the whole floor), how many halvings leave you with just 1 lentil?

Complete this table:

Number of Divisions	Fraction of Floor	Number of Lentils
0	1/1	16,284
1	1/2	8,192
2	1/4	|
3	1/8	(to 1 lentil)
↓	↓	↓

Halving Pattern

full floor:
$2^{14} = 16,384$

one-half floor:
$2^{13} = 8,192$

one-fourth floor:
$2^{12} = 4,096$

one-eighth floor
$2^{11} = 2,048$

one-sixteenth floor
$2^{10} = 1,024$

$2^9 = 512$

$2^7 = 128$

$2^8 = 256$

$2^5 = 32$

$2^6 = 64$

$2^4 = 16$

$2^3 = 8$

$= 2$

$= 1$

2^0 2^1

$2^2 = 4$

THIS WHOLE SHEET COVERS ONE-FOURTH FLOOR.

Preparation: 1 copy

TEACHING NOTES

Purpose

To estimate the number of lentils in a smaller volume, and extend that estimate to a larger volume.

Introduction

▸ This Cardboard Dam has the same area as the end of this Job Box. Where should we position it to confine all lentils to just 1/6 of the total volume? *Place it a distance of* 1/6 *from either end.*

(Other acceptable portions to dam are 1/8 or 1/10 of the Job Box. Because these fractions are smaller, the final estimate will be less accurate. Avoid fractions larger than 1/6; they require too many lentils to fill the reservoir.)

▸ How will you estimate the number of lentils required to fill the reservoir? *Track the number of quarts, tubs, pints, cups, etc, poured in; multiply by lentil numbers in the reference table; add it all together.* (This volume could also be measured in jam-packed liters and multiplied by 2^{14}.)

▸ How will you estimate the lentil capacity of your Job Box? Multiply the fraction by its denominator to equal the whole: i.e. $\frac{1}{6} \times 6 = 1$.

Focus

◆ What fraction of the Job Box will you dam up?
◆ How will you count the lentils?

Checkpoint

◆ What part of the Job Box did you dam up? Let me see your math.

◆ What is the total estimated lentil capacity of your Job Box? (We calculated our Job Boxes to hold between 200,000 and 230,000 lentils, a little over 1/5 million.)

Students may report answers like 213,456 lentils. Such a number has far too many significant figures. (Having gone through the work of multiplying, who wants to throw any figures away?!) Wonder aloud if the actual total might really be 213,45**7**, or more like 213,45**5**. This sounds pretty silly when imagining a box brimming with lentils. Could a Job Box hold 1 cup more or less and still look full? When estimating quantities like these, uncertainty happens in the *thousands* of lentils.

More

Check your estimate by damming up a different fraction of lentils. (The next Job Card presents an alternate way to estimate Job Box capacity and verify these answers.)

G/7 You need...

...the table of estimates you made in job card 4.

stick rulers

cardboard dam

up to 3 liters of lentils

DESIGN 3 rocks 3 cards

DIVIDE 11 containers

EMPTY job box

estimate 7

Block 1/6 of your Job Box with a Cardboard Dam held with rocks. Fill the reservoir with lentils.

How much does 1/6 of a job box hold?

Estimate how many lentils fill the **part**.

Estimate how many lentils fill the **whole**.

TEACHING NOTES

Purpose

To construct a one-inch cube and estimate how many lentils it holds. To use this information to calculate the volume and lentil capacity of the Job Box.

Introduction

Demonstrate how to cut and tape an inch cube. Line up the sides back-to-back by pinching them together. Then tape around the common edge.

Focus

◆ Why is this box called an inch cube? *Because it measures 1 inch by 1 inch by 1 inch.*

◆ How will you estimate the number of lentils it holds? *By finding out how many times I can fill it with lentils from the 1/6 cup, which holds ~500.*

◆ How will you estimate the number of lentils the Job Box holds? *Find its volume in cubic inches. Then multiply by the number of lentils per cubic inch.*

Checkpoint

◆ How many lentils does a cubic inch hold? *About 250. The 1/6 cup, estimated at 500 lentils, fills it fair and full twice.*

◆ How many inch cubes fill your Job Box? (Our particular box, measuring 14.5 x 19 x 3 inches, holds 825 cubic inches.)

◆ How many lentils fill your Job Box?
Sample calculation:
825 cu in x 250 lentils / cu in ≈ 206,000 lentils (As in the previous job card, this is about 1/5 million lentils.)

More

◆ Check the estimated capacity of your cubic inch with the Counting Grid. Repeat over several trials to establish a range of certainty.

◆ Estimate the volume of your Job Box again, using centimeters and/or lentils as the unit of measure. (A cubic centimeter holds about 15 lentils; a tiny cubic len holds 4. While these cubes are too small to be easily assembled from cross patterns, they can be successfully manipulated by young fingers when cut and taped without bottoms, as in the Job Card on page 56. These bottomless boxes are easy to fill as long as they rest on a flat surface.)

G/8 You need...

cross pattern

1 cubic inch

calculator

inch stick ruler

1/6 cup extra

job box with 1 liter of lentils

---

estimate 8   🯅 or 🯅🯅

Cut, fold and tape a cross pattern to make an inch cube.

1 cubic inch

INCH CUBE:

Estimate the number of lentils it holds.

1/6 cup

How many of these cubes fill your Job Box? How many lentils?

Cubic Inches = L x W x H

Compare your results with Job Card 7.

Cross Patterns

Divide into 6
squares
along the
dotted lines.

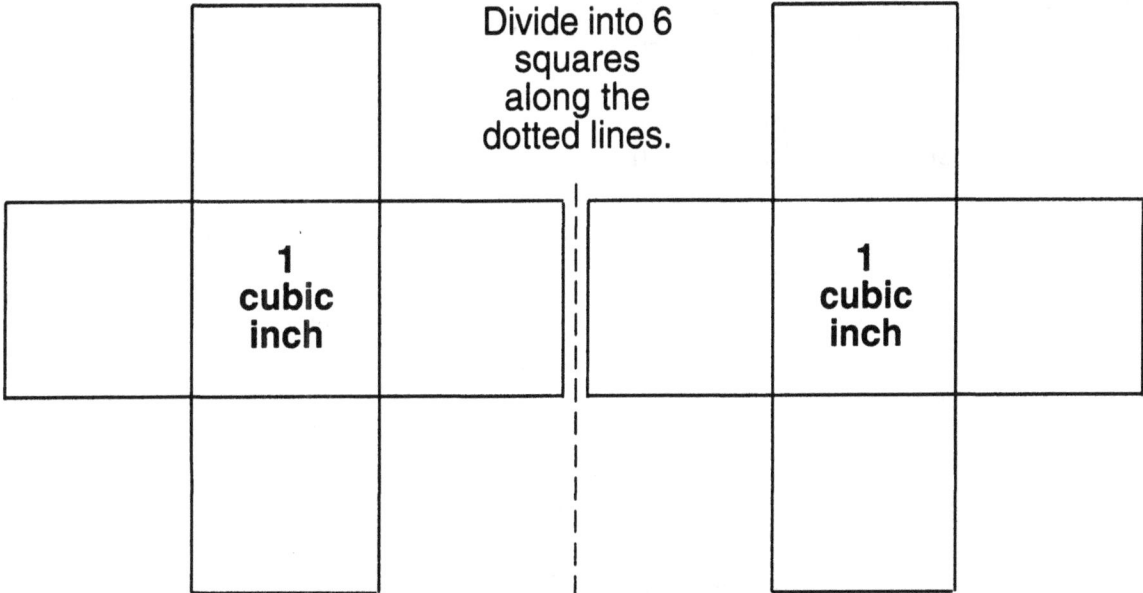

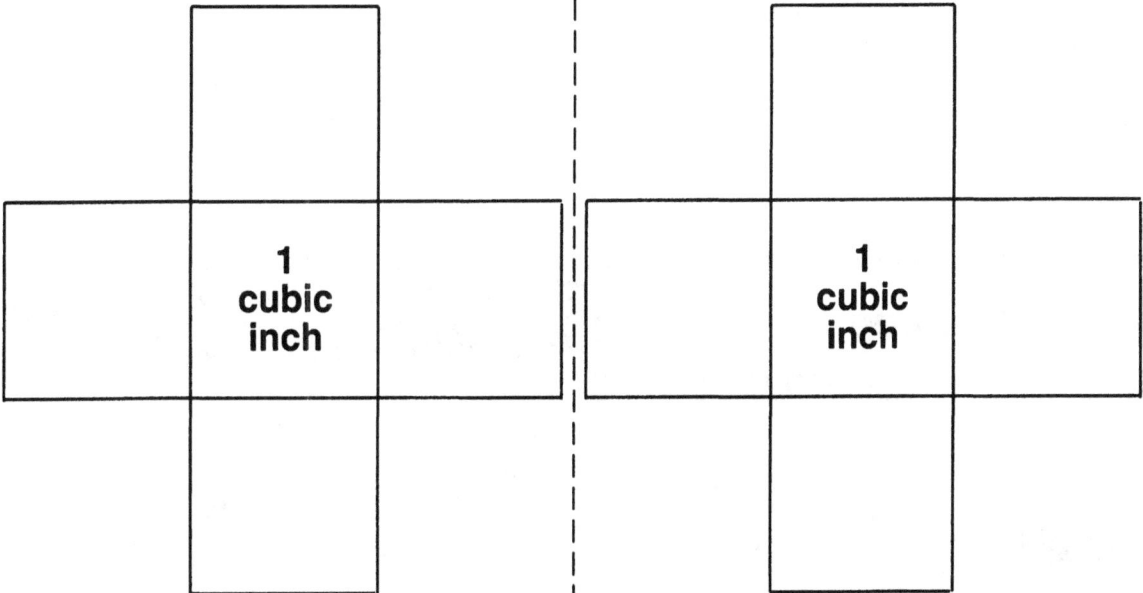

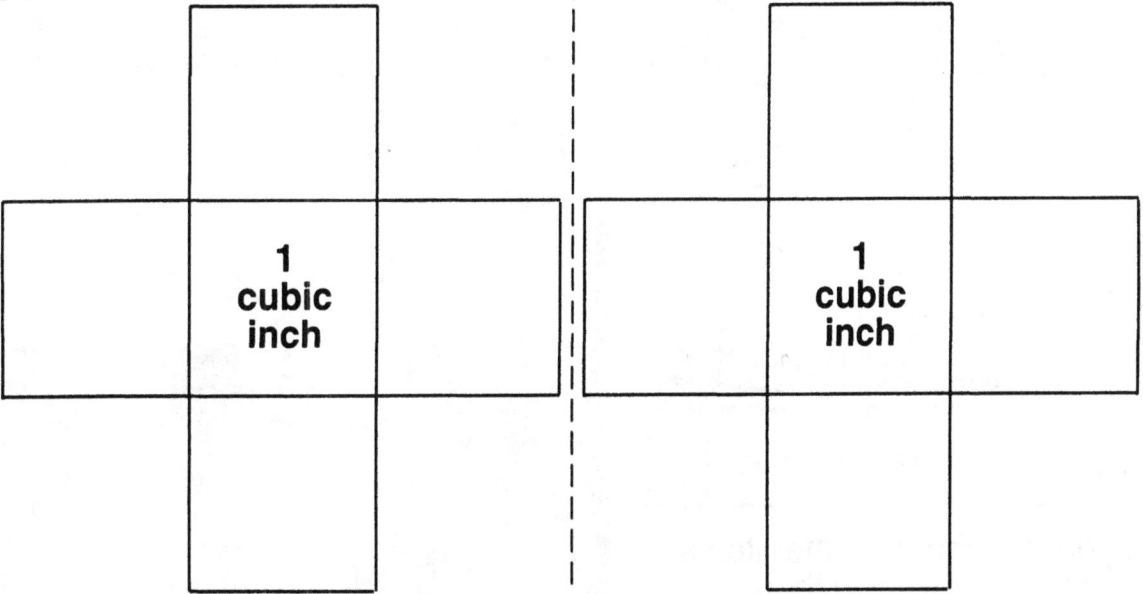

1
cubic
inch

1
cubic
inch

1
cubic
inch

1
cubic
inch

1
cubic
inch

1
cubic
inch

Preparation: several copies

TEACHING NOTES

Purpose

To provide students with an approved way to pursue their own ideas about estimating number.

Introduction

Display this card whenever you want to do this suggested activity or experiments that you design yourself.

Focus

◆ What do you want to study about estimating?

◆ Are these the materials you need?

Checkpoint

Report in words and pictures:
- What you did.
- What you learned.
- Questions you may still have.

Special Note

Tell TOPS about the most interesting new activities your students have designed. We may include them in future editions!

G/9 You may need...

cm stick

counting grid

spreader

calculator

scissors

DIVIDE 10 containers

short quart carton *(cut to hold 1 cup lentils)*

job box with **1** liter of lentils

☞ Ask your teacher for other items. Tell why you need them.

---

estimate 9 ♟ or ♟♟

On your own.

My idea:
Cut a one-cup box from a quart milk carton (7x7x4 cm). Since this box cup holds ≈ 3,000 lentils, I can use it to estimate how many lentils would fill my classroom!

Will I have to calculate the lentil volume of my body, too?

Feedback

If you enjoyed teaching TOPS please tell us so. Your praise motivates us to work hard. If you found an error or can suggest ways to improve this module, we need to hear about that too. Your criticism will help us improve our next new edition. Would you like information about our other publications? Ask us to send you our latest catalog free of charge.

For whatever reason, we'd love to hear from you. We include this self-mailer for your convenience.

Sincerely,

Ron and Peg Marson
author and illustrator

Your Message Here:

Module Title _____ Date _____

Name _____ School _____

Address _____

City _____ State _____ Zip _____

—————————————————————————————— FIRST FOLD ——————————————————————————————

—————————————————————————————— SECOND FOLD ——————————————————————————————

RETURN ADDRESS

_____

_____

_____

TOPS Learning Systems
342 S Plumas St
Willows, CA 95988

TAPE HERE

www.ingramcontent.com/pod-product-compliance
Lightning Source LLC
Chambersburg PA
CBHW080559220326
41599CB00032B/6544